NEW YORK STATE REGENTS
HIGH SCHOOL

GEOMETRY
INTEGRATED

ASSESSMENT EXAMINATION
REVIEW WORKBOOK

ANTHONY D'ONOFRIO
ANTHONY NIGRO
KEVIN O'REILLY
KATHLEEN THIBODEAU

ACKNOWLEDGMENTS
JOHN ALLASIO
EDWIN BERNAUER

Copyright © 2008

GEOMETRY WORKBOOK
ISBN-10: 0-937820-95-4
ISBN-13: 978-0-937820-95-7

TEACHERS' ANSWER KEY
ISBN-10: 0-937820-96-2
ISBN-13: 978-0-937820-96-4

NEW YORK STATE REGENTS HIGH SCHOOL

GEOMETRY

INTEGRATED

ASSESSMENT EXAMINATION REVIEW WORKBOOK

AUTHORS:

Anthony D'Onofrio
    Mathematics Department Chairperson (Retired)
    Co-author of Sequential Mathematics II Workbook
    Co-author of Regents High School Mathematics A Workbook
    Co-author of NYS Regents High School Algebra Workbook

Anthony Nigro
    Mathematics Teacher (Retired)
    Co-author of Sequential Mathematics I and III Workbooks
    Co-author of Regents High School Mathematics A and B Workbooks
    Co-author of New York State Intermediate Math 8 Workbooks, Vol. 1 & Vol. 2
    Co-author of NYS Regents High School Algebra Workbook

Kevin O'Reilly
    Mathematics Department Chairperson, Seaford UFSD
    Co-author of Regents High School Mathematics A Workbook
    Co-author of New York State Intermediate Math 8 Workbook, Vol. 1
    Co-author of NYS Regents High School Algebra Workbook

Kathleen Thibodeau
    Mathematics Department Chairperson (Retired)
    Co-author of Sequential Mathematics I Workbook
    Co-author of New York State Intermediate Math 8 Workbook, Vol. 2
    Co-author of NYS Regents High School Algebra Workbook

ACKNOWLEDGMENTS:

John Allasio
    Mathematics Department Chairperson (Retired)

Edwin Bernauer
    Mathematics Teacher (Retired)

www.WestseaPublishing.com   Copyright © 2008   WestSea Publishing Co. Inc.
                                                                                  149D Allen Boulevard
ISBN-10: 0-937820-95-4   Farmingdale, NY 11735-5616
ISBN-13: 978-0-937820-95-7
1.800.543.6130 (TOLL FREE)   **1.631.420.1110 (Phone)**   **1.631.420.0754 (FAX)**

INTRODUCTION

The Westsea Publishing Company is a forerunner in publishing review workbooks to meet the New York Learning Standards for Mathematics revised by the NYS Board of Regents March 15, 2005.

The Westsea <u>Regents High School Algebra Integrated Examination Review Workbook</u> was introduced at the AMTNYS Fall mathematics conference in 2006 for the first Algebra Integrated regents examination given June 2008. The algebraic skills and concepts within the Algebra process and content performance indicators must be maintained and applied by the students as they justify or prove geometric concepts.

Now Westsea is proud to introduce the <u>Regents High School Geometry Examination Review Workbook.</u> The first Geometry regent's examination, to be given June 2009, will test for student competence in the Geometric Strands.

The Geometric Strands include:
   <u>Geometric Relationships</u>.
   <u>Constructions</u> using a straightedge and compass.
   <u>Locus</u> simple and compound with graphing in the coordinate plane.
   <u>Informal and Formal Proofs</u>
   <u>Transformational Geometry</u>
   <u>Coordinate Geometry</u>

Educators using their own teaching and classroom techniques will find ample problems integrated in the above mentioned geometric strands that meet the requirements of the New York Learning Standards for Mathematics.

This <u>Regents High School Geometry Examination Review Workbook</u> also contains a chapter on using the Graphing Calculator. This chapter covers, step by step, the use of the TI-83 PLUS and TI-84 PLUS Graphing Calculators to solve problems similar to those solved by traditional methods in previous chapters.

In addition to a complete table of contents containing the strand numbers and strand names, there is also an index listing all the topics in alphabetical order affording easy access to a particular mathematical concept and its location in the workbook.

We at Westsea Publishing are proud of all our publications during the past 28 years and are confident that this latest <u>Regents High School Geometry Examination Review Workbook</u> will again contribute to student success.

The University of the State of New York
THE STATE EDUCATION DEPARTMENT
Albany, New York 12234

Specifications for the Regents Examination in Geometry
(First Administration–June 2009)

The questions on the Regents Examination in Geometry will assess both the content and the process strands of New York State Mathematics Standard 3. Each question will be aligned to one content performance indicator but will also be aligned to one or more process performance indicators, as appropriate for the concepts embodied in the task. As a result of the alignment to both content and process strands, the examination will assess students' conceptual understanding, procedural fluency, and problem-solving abilities rather than assessing knowledge of isolated skills and facts.

There will be 38 questions on the Regents Examination in Geometry. The table below shows the percentage of total credits that will be aligned with each content band.

Content Band	% of Total Credits
Geometric Relationships	8–12%
Constructions	3–7%
Locus	4–8%
Informal and Formal Proofs	41–47%
Transformational Geometry	8–13%
Coordinate Geometry	23–28%

Question Types

The Regents Examination in Geometry will include the following types and numbers of questions:

Question Type	Number of Questions
Multiple choice	28
2-credit open ended	6
4-credit open ended	3
6-credit open ended	1
Total credits	86

Calculators

Schools must make a graphing calculator available for the exclusive use of each student while that student takes the Regents Examination in Geometry.

REGENTS HIGH SCHOOL
GEOMETRY
EXAM REVIEW WORKBOOK

TABLE OF CONTENTS

CHAPTER 1

LOGIC (INFORMAL PROOFS)   **PAGE**

G.G.24. Negation of a statement and Truth Value 1, 2
G.G.25. Negation of a conditional statement... 3
G.G.25. Negation of "ALL" or "SOME" statements.. 4
G.G.25. Conjunctions, conditions when true 5, 6
G.G.25. Disjunctions, conditions when true 7, 8
G.G.25. Conditional, conditions when true 9
G.G.25. Biconditional, conditions when true 10
G.G.26. Inverse of a conditional statement... 11
G.G.26. Converse of a conditional statement 12
G.G.26. Contrapositive of a conditional statement.. 13
G.G.26. Completing truth tables 14
G.G.26. Determining tautologies 14, 15

CHAPTER 2

GEOMETRIC PROOFS (INFORMAL & FORMAL)

G.G.28. Congruence of triangles, SSS, SAS, ASA, AAS, HYP-LEG.. 16-18
G.G.38. Parallelograms, involving angles, sides, diagonals 19-20
G.G.39. Rectangles, involving angles, sides, diagonals 21
G.G.39. Rhombuses, involving angles, sides, diagonals 22
G.G.39. Squares, involving angles, sides, diagonals 23
G.G.40. Trapezoids, involving angles, sides, medians, diagonals... 24
G.G.40. Trapezoids, isosceles, involving angles, sides, diagonals 24
G.G.27. Formal proofs, congruence, hypothesis to conclusion 25-29
G.G.29. Corresponding parts of congruent triangles, CPCTC 25-29
G.G.27. Formal proofs, similarity, hypothesis to conclusion 30, 31
G.G.44. Similarity of triangles, AA 30, 31
G.G.45. Similar triangles, corresponding sides are in proportion 32-35
G.G.45. Similar triangles, product of means = product of extremes 32-35

CHAPTER 3

MEASUREMENT IN POLYGONS PAGE

G.G.30. The sum of the angles of a triangle 36
G.G.32. Exterior angle theorem of a triangle.. 3 -38
G.G.31. The bisector of the vertex angle of an isosceles triangle... 39
G.G.33., G.G.34. Inequality theorem of a triangle 40-42
G.G.35. Parallel and intersecting lines... 43-45
G.G.36. Sum of interior and exterior angles in polygons 46
G.G.37. Each interior and exterior angle of a regular polygon... 47
G.G.38. Parallelograms... 48
G.G.39., G.G.41. Special parallelograms 49
G.G.42., G.G.43. Midpoints and centroids in triangles 50
G.G.47. Right triangle proportions 51, 52
G.G.46. Proportional sides of a triangle 53
G.G.48. Pythagorean theorem.. 54, 55
G.G.40. Trapezoids 56, 57

CHAPTER 4

MEASUREMENT IN CIRCLES

G.G.49. Chords, arcs and perpendicular bisectors... 58, 59
G.G.51. Central and inscribed angles 60
G.G.51. Angles formed by a tangent and a chord 61
G.G.51. Angles formed by two chords 62
G.G.51. Angles formed by two tangents or two secants.. 63
G.G.51. Angles formed by a tangent and a secant 63
G.G.51. Angles in circles, combination 64, 65
G.G.53. Lengths of tangent and secants 66
G.G.53. Lengths of two intersecting chords 67
G.G.53. Lengths of two secants 68
G.G.50., G.G.52. Parallel and perpendicular lines in circles... 69
G.G.50. Common external and internal tangents.. 70, 71

CHAPTER 5

COORDINATE GEOMETRY   PAGE

G.G.62. Slope of a line perpendicular to a given equation of a line 72
G.G.63. Determine if two lines are parallel, perpendicular, or neither 73
G.G.64, G.G.65 Given a point on the line and the equation of
G.G.64. a line perpendicular to the given line 74
G.G.65. a line parallel to the given line 75
G.G.66. Midpoint of a line segment, given its endpoints 76
G.G.67. Length of a line segment, given its endpoints 77
G.G.68. Equation of a line that is the perpendicular bisector 78
G.G.69. Properties of triangles using distance, midpoint, and slope 79, 81
G.G.69. Properties of quadrilaterals using distance, midpoint, slope 80, 83
G.G.70. Graphing systems of equations, one linear and one quadratic 87
G.G.71. Equation of a circle, given its center and radius 88
G.G.71 Equation of a circle, given its endpoints of its diameter 88
G.G.72 Equation of a circle, given its graph 89
G.G.73 Equation of a circle, find center and radius 90
G.G.74 Graph circles of the form $(x - h)^2 + (y - k)^2 = r^2$ 91-93
G.G.27e Prove the diagonals of a rectangle are congruent 94, 95
G.G.43a Determine the coordinates of the centroid of the triangle 96
G.G.43a Find the lengths of the segments of the medians 97, 98

CHAPTER 6

TRANSFORMATIONAL GEOMETRY

G.G.54 Translations 99-101
G.G.54 Line reflections 102-105
G.G.54 Glide reflections 106-108
G.G.54 Rotations 109-111
G.G.61 Dilations 112-115
G.G.58 Composition of isometrics and dilations 116-122
G.G.57 Isometries in the plane 123-126
G.G.56 Identifying specific isometries 127-137
G.G.58 Identifying specific similarities 138-141
G.G.55 Properties that remain invariant under isometries and similarities 142-145
G.G.59 Properties that remain invariant under similarities 146-148

CHAPTER 7

LOCUS PAGE

G.G.21 Locus, simple 149-152
G.G.22 Locus, compound 153, 154
G.G.23 Locus, graphing compound loci in the coordinate plane 155-157

CHAPTER 8

CONSTRUCTIONS

G.G.17 Constructions, bisector of a given angle 158-161
G.G.18 Constructions, perpendicular bisector of a segment 158-160
G.G.19 Constructions, lines parallel or perpendicular 158-161
G.G.20 Constructions, equilateral triangle 159, 160
G.G.18 Constructions, circumcenter of a triangle 162
G.G.17 Constructions, incenter of a triangle 162
G.G.18 Constructions, centroid of a triangle 162
G.G.19 Constructions, orthocenter of a triangle 163

CHAPTER 9

GEOMETRIC RELATIONSHIPS

G.G.1 Line perpendicular to two intersecting lines 164
G.G.2 Plane perpendicular to a given line 165
G.G.3 Line perpendicular to a given plane, through a given point 166
G.G.4 Coplanar lines 167
G.G.5 Two planes perpendicular to each other 168
G.G.6 Line perpendicular to a given plane, line perpendicular 169
G.G.7 Line perpendicular to a given plane, plane contains 170
G.G.8 Plane intersecting two parallel planes 171
G.G.9 Two planes perpendicular to same line 172
G.G.10 Lateral edges of a prism 173
G.G.11 Prisms with equal volumes 174, 175
G.G.12 Prism, volume formula 176-182
G.G.13, G.G.48b Regular pyramid properties 183-189
G.G.14 Cylinder properties 190-195
G.G.15 Right circular cone properties 196-200
G.G.16 Sphere properties 201-207
G.G.48a Pythagorean theorem, rectangular solids 208-210
G.G.14, G.G.15 Ratios in similar right circular cylinders & cones 211-212

CHAPTER 10

GRAPHING CALCULATOR   PAGE

Solving equations using the TI-83 Plus or TI-84 Plus "SOLVER" 213
Solving equations by determining the zeros of a function... 214
Graphing polygons using scatterplota... 215
Using scatterplots and lists to transform polygons 216, 217
Using scatterplots and lists to rotate polygons about the origin 218, 219
Using scatterplots and lists to reflect polygons over a line 220, 221
Graphing circles 222, 223
Conditions under which a compound sentence is true... 224
Using geometry software to make conjectures 225-230

REFERENCE SHEET

Regents Examination in Geometry formula sheet 231

INDEX 232-236

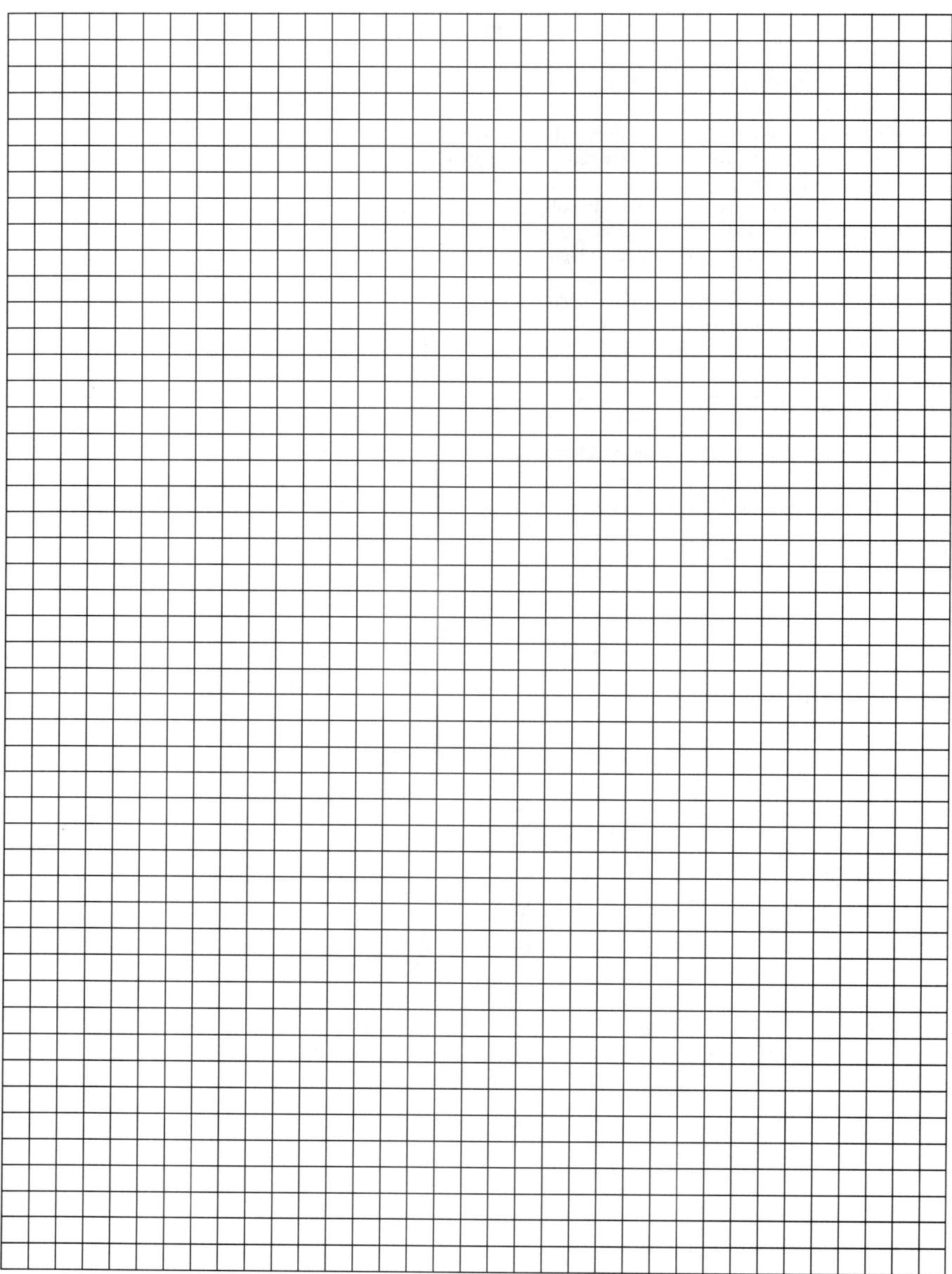

x

Chapter 1

LOGIC
Determine the Negation of a Statement and Establish its Truth Value

REMEMBER

A STATEMENT is a simple sentence with a TRUTH VALUE of either TRUE or FALSE.

To NEGATE a statement, place the word "NOT" in the statement.

Symbolically, the letter p often represents a statement and $\sim p$ (not p) represents the negation of the statement. The symbol $\sim$ is known as a "tilde".

Example: What is the negation of the statement "The capital of New York State is New York City." Is the negation TRUE or FALSE?

Answer: The capital of New York State is not New York City.
Since the capital of New York State is Albany, this new statement is true.

1.	Write the negation of the statement "The Earth revolves around the Sun". Is the negation TRUE or FALSE?	5.	Write the negation of the statement "The solution to the equation $3(x-4)+5=2$ is $x=1$". Is this negation TRUE or FALSE?
2.	Write the negation of the statement "There are nine planets that revolve around the Sun". Is the negation TRUE or FALSE?	6.	Write the negation of the statement "Triangles can be proved congruent using the Side-Side-Angle theorem". Is this negation TRUE or FALSE?
3.	Write the negation of the statement "The square root of 16 is always 4". Is this negation TRUE or FALSE?	7.	Write the negation of the statement "A conditional statement and its inverse always have the same truth value". Is this negation TRUE or FALSE?
4.	Write the negation of the statement "Volume is a three-dimensional concept". Is this negation TRUE or FALSE?	8.	Write the negation of the statement "A conditional statement and its contrapositive do not have the same truth value". Is this negation TRUE or FALSE?

Chapter 1

LOGIC
Determine the Negation of a Statement and Establish its Truth Value (continued)

9.	Write the negation of the statement "A person on trial is always guilty". Is this negation TRUE or FALSE?	13.	Write the negation of the statement "A diagonal of an isosceles trapezoid divides the trapezoid into two congruent triangles". Determine the truth value of both the statement and its negation.
10.	Which of the following is the negation of the statement "The best team will win the game"? (1) The best team will lose the game. (2) The game will end in a tie. (3) Another team will win the game. (4) The best team will not win the game.	14.	If the letter p represents the statement "I will pass the test", which of the following represents the statement "I will not pass the test"? (1) p (3) $\sim p$ (2) $\approx p$ (4) $\neq p$
11.	Write the negation of the statement "A square is a rhombus". Determine the truth value of both the statement and its negation.	15.	If p represents a true statement, which of the following also represents a true statement? (1) $\sim p$ (3) $\sim(\sim(\sim p))$ (2) $\sim(\sim p)$ (4) $\neq p$
12.	Write the negation of the statement "A quadrilateral with perpendicular diagonals is a rhombus". Determine the truth value of both the statement and its negation.	16.	If $\sim p$ represents a true statement, which of the following also represents a true statement? (1) p (3) $\sim(\sim(\sim p))$ (2) $\sim(\sim p)$ (4) $\sim(\sim(\sim(\sim p)))$

Chapter 1

LOGIC
Determine the Negation of a Conditional Statement

REMEMBER

A conditional statement is one in the form of "IF ... THEN ..." such as "If there is water on Mars, then there is life on Mars."

To negate a conditional statement, put the phrase "It is not true that" in front of the statement. The negation of the statement above is "It is not true that if there is water on Mars then there is life on Mars". Do not negate each part of the statement.

A conditional statement can be represented symbolically as $p \to q$ where p represents the hypothesis and q represents the conclusion. Read $p \to q$ as "If p, then q" or "p implies q".

To negate the conditional statement $p \to q$, write $\sim(p \to q)$. Do not negate p and q separately.

---

1. Write the negation of the following statement:

   "If Meghan studies, then she will get an A."

2. Write the negation of the following statement:

   "If Aidan finishes his dinner, then he can have ice cream for dessert."

3. Write the negation of the given statement. Determine the truth value of both the statement and its negation.

   "If two lines are cut by a transversal, then the corresponding exterior angles are congruent."

4. Write the negation of the given statement. Determine the truth value of both the statement and its negation.

   " If the quadrilateral has four congruent sides then it is a rhombus."

5. Write the negation of the given statement. Determine the truth value of both the statement and its negation.

   "If a quadrilateral is a parallelogram, then its diagonals are congruent."

6. Which of the following represents the negation of the conditional statement $p \to q$?

   (1) $\sim p \to q$ (3) $\sim p \to \sim q$
   (2) $p \to \sim q$ (4) $\sim(p \to q)$

Chapter 1

LOGIC
Determine the Negation of All or Some Statements

> **REMEMBER**
>
> A statement using the word ALL or EVERY is known as a universal quantifier. To negate an ALL or EVERY statement, use either the word SOME or the phrase THERE EXISTS AT LEAST ONE and negate the given statement.
>
> EXAMPLE: Negate the phrase "All mathematics teachers use chalk".
> ANSWER: "Some mathematics teachers do not use chalk."
>
> A statement using the word SOME is an existential quantifier. To negate a SOME statement use the word ALL or EVERY and negate the given statement.
>
> EXAMPLE: Negate the phrase "Some triangles are isosceles".
> ANSWER: "All triangles are not isosceles".

1. Negate the statement:

   "Every student has their textbook."

2. Negate the statement:

   "All music is fun to dance to."

3. Negate the given statement. Determine the truth value of the statement and its negation.

   "Every rhombus is a square."

4. Negate the given statement. Determine the truth value of the statement and its negation.

   "All radii for a given circle are the same length."

5. Negate the statement:

   "Some trees grow over 100 feet tall."

6. Negate the statement:

   "Some mathematics courses are not easy."

7. Negate the given statement. Determine the truth value of the statement and its negation.

   "Some triangles are not equilateral."

8. Negate the given statement. Determine the truth value of the statement and its negation.

   "There is a parallelogram that is a rectangle."

Chapter 1

LOGIC
Know and Apply the Conditions Under Which a Conjunction is True

REMEMBER

A **conjunction** is a compound sentence formed by combining two simple sentences with the word "and". A **conjunction** is **true** only when <u>**both simple sentences are true**</u>.

Symbolically, the symbol ∧ represents the word "and".

Example: Symbolically write the sentence "I am going to New York City and I am visiting President Grant's Tomb" if p represents the sentence "I am going to New York City" and if q represents the sentence "I am visiting President Grant's Tomb". When will the given compound sentence be true?

Answer: $p \wedge q$. The sentence "I am going to New York City and I am visiting President Grant's Tomb" will only be true if I actually go to NYC and visit Grant's Tomb.

---

1. Consider two events A and B represented by the two circles in the Venn Diagram below. Shade the region that represents events A and B both occurring.

   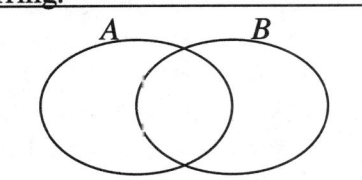

2. When $x = 3$, which of the following is true?

   (1) x is prime and x is odd.
   (2) x is prime and x is even.
   (3) x is the smallest prime number and x is odd.
   (4) $x + 1$ is prime and $x + 1$ is odd.

3. If Helen owns a cat is true, what is the truth value of "Helen owns a cat and a dog"?

   (1) True   (3) Helen owns a cat.
   (2) False   (4) Can not be determined.

4. Complete the truth table below for the value of $p \wedge q$ given the truth values of the simple sentences p and q.

p	q	$p \wedge q$
TRUE	TRUE	
TRUE	FALSE	
FALSE	TRUE	
FALSE	FALSE	

5. If the statement "All dogs can not fly and some fish can fly" is true, which of the following statements must be true?

   (1) Some dogs can fly.
   (2) All dogs can fly.
   (3) All fish can fly.
   (4) Some fish can fly.

6. A restaurant offers a dinner with a choice of Soup or Salad, and Steak or Chicken, and Potato or Corn. Which of the following dinners may be chosen?

   (1) Soup, Salad and Steak
   (2) Salad, Chicken and Potato
   (3) Soup, Steak, Potato and Corn
   (4) Salad, Potato and Corn

Chapter 1

LOGIC
Know and Apply the Conditions Under Which a Conjunction is True (continued)

7. The statement "x is divisible by 2 and x is divisible by 3" is false if x equals (1) 6 (3) 15 (2) 12 (4) 18	12. The statement "$x \leq 2$ and $x \geq 4$" is true if x equals (1) The statement is never true. (2) 2 (3) 3 (4) 4
8. The statement "x is prime and x is even" is true if x equals (1) 1 (3) 3 (2) 2 (4) 4	13. For which of the following values is the statement "$x \geq 5$ and $x \leq 7$" false? (1) 5 (3) 7 (2) 6 (4) 8
9. For which value of x is the statement "x is even and $\sqrt{x}$ is prime" true? (1) 2 (3) 9 (2) 4 (4) 16	14. From the set {parallelogram, rectangle, rhombus, square, trapezoid}, how many of the figures make the statement "My opposite sides are parallel and my diagonals have the same length" false? (1) 1 (3) 3 (2) 2 (4) 4
10. Determine the positive integers less than 100 that make the statement "I'm a perfect square and I'm a perfect cube" true.	15. Which of the following statements is true for all rhombuses? (1) All sides are equal in length and diagonals are equal in length. (2) Consecutive angles are equal in measure and diagonals are perpendicular. (3) Diagonals are equal in length and all vertex angles measure 90°. (4) Diagonals are perpendicular and consecutive angles are supplementary.
11. For which type of quadrilateral is the statement "My opposite sides are parallel and my diagonals are the same length" true? (1) Parallelogram (3) Rhombus (2) Isosceles Trapezoid (4) Rectangle	

Chapter 1

LOGIC
Know and Apply the Conditions Under Which a Disjunction is True

REMEMBER

A **disjunction** is a compound sentence formed by combining two simple sentences with the word "or". A **disjunction** is **false** only when **both simple sentences are false**.

Symbolically, the symbol ∨ represents the word "or".

Example: Symbolically write the sentence "I will eat ice cream or I will eat cake" if p represents the sentence "I will eat ice cream" and if q represents the sentence "I will eat cake". When will the given compound sentence be true?

Answer: $p \vee q$. The sentence "I will eat ice cream or I will eat cake" will only be true if I actually eat ice cream, cake or both.

---

1. Consider two events A and B represented by the two circles in the Venn Diagram below. Shade the region that represents event A or B occurring.

   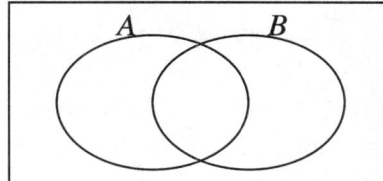

2. When $x = 2$, which of the following is false?

   (1) x is prime or x is even.
   (2) x is prime or x is odd.
   (3) x is the smallest prime number or x is odd.
   (4) $x + 2$ is prime or x^2 is odd.

   _____

3. If Helen owns a cat is true, what is the truth value of "Helen owns a cat or a dog"?

   (3) True   (3) Helen owns a cat.
   (2) False   (4) Can not be determined.

   _____

4. Complete the truth table below for the value of $p \vee q$ given the truth values of the simple sentences p and q.

p	q	$p \vee q$
TRUE	TRUE	
TRUE	FALSE	
FALSE	TRUE	
FALSE	FALSE	

5. An above average score on the Writing, Mathematics or Critical Reading portion of the SAT is required for acceptance to a college.

   If the average score for Writing is 497, Mathematics is 517 and Critical Reading is 505 what is the minimum score (rounded to the nearest 10) and on which section of the SAT is necessary to be accepted to this college? Explain your answer.

   _____

Chapter 1

LOGIC

Know and Apply the Conditions Under Which a Disjunction is True (continued)

7. The statement "x is divisible by 2 or x is prime" is false if x equals

    (1) 1   (3) 3
    (2) 2   (4) 4

8. The statement "x is a perfect square or x is prime" is true if x equals

    (1) 10   (3) 6
    (2) 8   (4) 4

9. For which value of x is the statement "x is not even or $\sqrt{x}$ is prime" true?

    (1) 0   (3) 8
    (2) 2   (4) 4

10. Determine the positive integers less than 100 that make the statement "I'm a perfect square or I'm a perfect cube" true.

11. For which type of quadrilateral could the statement "My opposite sides are parallel or my diagonals are the same length" be false?

    (1) Parallelogram   (3) Rhombus
    (2) Trapezoid   (4) Rectangle

12. The statement "$x < 2$ or $x > 4$" is true if x equals

    (1) 1   (3) 3
    (2) 2   (4) 4

13. For which of the following values is the statement "$x \geq 5$ or $x \leq 7$" false?

    (1) 5   (3) 7
    (2) 6   (4) The statement is always true.

14. From the set {parallelogram, rectangle, rhombus, square, isosceles trapezoid}, how many of the figures make the statement "My opposite sides have the same length or my diagonals have the same length" true?

    (1) 5   (3) 3
    (2) 2   (4) 4

15. Which of the following statements is true for all parallelograms?

    (1) All sides are equal in length or diagonals are equal in length.
    (2) Consecutive angles are equal in measure or diagonals are perpendicular.
    (3) Diagonals are equal in length or all vertex angles measure 90°.
    (4) Diagonals are perpendicular or consecutive angles are supplementary.

Chapter 1

LOGIC
Know and Apply the Conditions Under Which a Conditional is True

> **REMEMBER**
>
> A **conditional** is a compound sentence formed by beginning a one simple sentence with **"if"** (the hypothesis) and the other simple sentence with **"then"** (the conclusion). A **conditional** is **false** only when the **"if" sentence is true** and the **"then" sentence is false.**
>
> Symbolically, the symbol $\rightarrow$ represents the conditional sentence "If ... then ..."
>
> **Example:** Symbolically write the sentence "If I do my homework, then I will pass the course" if p represents the sentence "I do my homework" and if q represents the sentence "I will pass the course". When will the given compound sentence be false?
>
> **Answer:** $p \rightarrow q$. The sentence "If I do my homework, then I will pass the course" will only be false if I actually do my homework and do not pass the course.

1. For which value of x is the statement "If x is odd, then x is prime" false?

    (1) 1
    (2) 2
    (3) 3
    (4) 4

    _____

2. For which figure is the statement "If the quadrilateral is a parallelogram, then the diagonals are congruent" false?

    (1) rectangle
    (2) trapezoid
    (3) square
    (4) rhombus

    _____

3. When is the statement "If Thomas goes to the game, he will catch a foul ball" false?

    (1) Thomas does not go to the game and thus did not catch a foul ball.
    (2) Thomas does not go to the game, but his friend catches a foul ball.
    (3) Thomas goes to the game and does not catch a foul ball.
    (4) Thomas goes to the game and he catches a foul ball.

4. Complete the truth table below for the value of $p \rightarrow q$ given the truth values of the simple sentences p and q.

p	q	$p \rightarrow q$
TRUE	TRUE	
TRUE	FALSE	
FALSE	TRUE	
FALSE	FALSE	

5. Consider the statement "If Jay go to the movies, then Jay will have a good time" to be true. If Jay does go to the movies, which of the following must be correct?

    (1) Jay will have a good time.
    (2) Jay will not have a good time.
    (3) Jay will purchase popcorn.
    (4) Jay will not like the movie.

    _____

6. Consider the true statement "If a triangle is isosceles than exactly two angles are congruent". If a triangle does not have exactly two congruent angles, which of the following must be true?

    (1) The triangle is isosceles.
    (2) The triangle is not isosceles.
    (3) The triangle is equilateral.
    (4) The triangle is scalene.

    _____

Chapter 1

LOGIC
Know and Apply the Conditions Under Which a Biconditional is True

REMEMBER

A **biconditional** is a compound sentence formed by combining two simple sentences with the words "if and only if". A **biconditional** is **true** whenever **both simple statements are the same** (*both true or both false*).

Symbolically, the symbol $\leftrightarrow$ represents the biconditional statement and is read as "… if and only if …" A biconditional statement can be rewritten as a conditional statement and its converse. A biconditional statement is true if both the conditional statement and its converse are true.

Example: Symbolically write the sentence "$x + 10 = 17$ if and only if $x = 7$" if p represents the sentence "$x + 10 = 17$" and if q represents the sentence "$x = 7$". Rewrite the statement as a conditional statement and its converse.

Answer: $p \leftrightarrow q$. If $x + 10 = 17$ then $x = 7$ and if $x = 7$ then $x + 10 = 17$.

---

1. Determine whether the statement below is true or false:

    "$\sqrt{17} = 4$ if and only if the Earth is smaller than the moon."

    (1) True (3) Can not be determined.
    (2) False (4) The moon is made of cheese.

2. Determine whether the statement below is true or false:

    "$x^2 > 0$ if and only if x is not equal to zero."

    (1) True (3) Can not be determined.
    (2) False (4) x^2 must be greater than zero.

3. Which of the following numbers satisfies the statement "I am a perfect square if and only if I am a perfect cube"?

    (1) 4   (3) 16
    (2) 8   (4) 64

4. Complete the truth table below for the value of $p \leftrightarrow q$ given the truth values of the simple sentences p and q.

p	q	$p \leftrightarrow q$
TRUE	TRUE	
TRUE	FALSE	
FALSE	TRUE	
FALSE	FALSE	

5. Which of these biconditionals is false?

    (1) A triangle is equilateral if and only if each angle measures 60°.
    (2) A triangle is isosceles if and only if exactly two of its angles are congruent.
    (3) A triangle is a right triangle if and only if one angle measures 90°.
    (4) A triangle is obtuse if and only if two of its angles are acute.

6. Which of these biconditionals is false?

    (1) $x^2 = 0$ if and only if $x = 0$
    (2) $x^2 \geq 0$ if and only if $x \geq 0$
    (3) $x^2 > 0$ if and only if $|x| > 0$
    (4) $x^2 \geq 0$ of and only if $|x| \geq 0$

Chapter 1

LOGIC

Identify and Write the Inverse of a Statement

> **REMEMBER**
>
> The inverse of the conditional statement $p \to q$ is determined by negating both the hypothesis (p) and the conclusion (q) to obtain $\sim p \to \sim q$.
>
> A conditional statement and its inverse are not logically equivalent.
>
> Example: What is the inverse of the statement "If I study, then I will not fail this course".
>
> Answer: "If I do not study, then I will fail this course."

1. Write the inverse of the statement "If you have more than 17 absences in a class, you will be denied credit for the course".

2. Write the inverse of the statement "If you break a school rule at the senior prom, then you will not be allowed at graduation".

3. Write the inverse of the statement below and determine if the statement and/or its inverse are true.

   "If a triangle is isosceles, then its altitude from the vertex angle bisects the base."

4. Write the inverse of the statement below and determine if the statement and/or its inverse are true.

   "If the diagonals of a quadrilateral are perpendicular, the quadrilateral is a rhombus."

5. Complete the truth table below to show that a conditional statement and its inverse are not logically equivalent:

p	q	$\sim p$	$\sim q$	$p \to q$	$\sim p \to \sim q$
T	T				
T	F				
F	T				
F	F				

6. Which of the following represents the inverse of $\sim p \to q$?

   (1) $\sim p \to \sim q$ (3) $q \to \sim p$
   (2) $p \to \sim q$ (4) $\sim q \to p$

   _____

7. Write the inverse of the statement below and determine if the statement and/or its inverse are true.

   "If a radius of a circle passes through the midpoint of a chord, then the radius is perpendicular to that chord."

8. Write the inverse of the statement below and determine if the statement and/or its inverse are true.

   "If a transversal falling on two straight lines makes the interior angles on the same side less than 180°, then the two lines are not parallel."

Chapter 1

LOGIC
Identify and Write the Converse of a Statement

> **REMEMBER**
>
> The converse of the conditional statement $p \rightarrow q$ is determined by switching both the hypothesis (p) and the conclusion (q) to obtain $q \rightarrow p$.
>
> A conditional statement and its converse are not logically equivalent.
>
> Example: What is the inverse of the statement "If I study, then I will not fail this course".
>
> Answer: "If I do not fail this course that I studied."

1. Write the converse of the statement "If you clean your room then you will not have to mow the lawn".

2. Write the converse of the statement "If the train is not on time to the station, then the train must have been traveling too slowly."

3. Write the converse of the statement below and determine if the statement and/or its inverse are true.

   "If $a^2 + b^2 \neq c^2$, then $\triangle ABC$ is not a right triangle."

4. Write the converse of the statement below and determine if the statement and/or its inverse are true.

   "If two planes are perpendicular to the same line, then the two planes are parallel".

5. Complete the truth table below to show that a conditional statement and its converse are not logically equivalent:

p	q	$p \rightarrow q$	$q \rightarrow p$
T	T		
T	F		
F	T		
F	F		

6. Which of the following represents the converse of $p \rightarrow \sim q$?

   (1) $\sim p \rightarrow \sim q$ (3) $q \rightarrow \sim p$
   (2) $p \rightarrow \sim q$ (4) $\sim q \rightarrow p$

   _____

7. Write the converse of the statement below and determine if the statement and/or its inverse are true.

   "If two lines are parallel then their slopes are equal."

8. Write the converse of the statement below and determine if the statement and/or its inverse are true.

   "If the diagonals of a quadrilateral are congruent, then each angle of the quadrilateral measures 90°."

Chapter 1

LOGIC
Identify and Write the Contrapositive of a Statement

> **REMEMBER**
>
> The contrapositive of the conditional statement $p \rightarrow q$ is determined by switching and negating both the hypothesis (p) and the conclusion (q) to obtain $\sim q \rightarrow \sim p$.
>
> A conditional statement and its contrapositive are always logically equivalent.
>
> Example: What is the contrapositive of the statement "If I study, then I will not fail this course".
>
> Answer: "If I fail this course that I did not study."

1. Write the contrapositive of the statement "If I am tired, then I will go to sleep".

2. Write the contrapositive of the statement "If Lisa sells the house, then her income will not remain the same".

3. Write the contrapositive of the statement below and determine if the statement and its' inverse are true. If the statement is not true, correct it.

   "If a quadrilateral's diagonals bisect each other, then the quadrilateral is a parallelogram."

4. Write the contrapositive of the statement below and determine if the statement and its' inverse are true. If the statement is not true, correct it.

   "If a quadrilaterals diagonals are perpendicular, then the quadrilateral is a square."

5. Complete the truth table below to show that a conditional statement and its contrapositive are logically equivalent:

p	q	$\sim p$	$\sim q$	$p \rightarrow q$	$\sim q \rightarrow \sim p$
T	T				
T	F				
F	T				
F	F				

6. Which of the following represents the contrapositive of $p \rightarrow \sim q$?

   (1) $\sim p \rightarrow \sim q$ (3) $q \rightarrow \sim p$
   (2) $p \rightarrow \sim q$ (4) $\sim q \rightarrow p$

   _____

7. Write the contrapositive of the statement below and determine if the statement and its' inverse are true. If the statement is not true, correct it.

   "If the exterior angle of a regular polygon is at most 45°, then the polygon has at least 6 sides."

8. Write the contrapositive of the statement below and determine if the statement and its' inverse are true. If the statement is not true, correct it.

   "If two triangles have equal bases and equal heights, then the area of the two triangles is the same."

Chapter 1

LOGIC
Completing Truth Tables and Determining Tautologies

REMEMBER

A **statement** is a simple sentence with that has a truth value of either **TRUE** or **FALSE**.

Statements which use the connectives "not", "and", "or", "if . . . then . . ." and/or ". . . if and only if . . ." are called **compound statements**.

A **tautology** is a compound statement whose truth value is always **TRUE** regardless of the truth value of the simple statements that compose the compound statement.

Two statements are **logically equivalent** if they form a tautology when connected with a biconditional.

A **contradiction** is a compound statement whose truth value is always **FALSE** regardless of the truth value of the simple statements that compose the compound statement.

A **contingency** is a compound statement whose truth value could be either **TRUE** or **FALSE** depending on the truth value of the simple statements that compose the compound statement.

Example: Determine if the statement $\sim(p \vee q) \leftrightarrow \sim p \wedge \sim q$ is a tautology.

Answer: Create a truth table for the statement by making columns that list all the possible truth value combinations for the sentences that combine to make the statement $\sim(p \vee q) \leftrightarrow \sim p \wedge \sim q$.

p	q	$\sim p$	$\sim q$	$p \vee q$	$\sim(p \vee q)$	$\sim p \wedge \sim q$	$\sim(p \vee q) \leftrightarrow \sim p \wedge \sim q$
T	T	F	F	T	F	F	T
T	F	F	T	T	F	F	T
F	T	T	F	T	F	F	T
F	F	T	T	F	T	T	T

The statement $\sim(p \vee q) \leftrightarrow \sim p \wedge \sim q$ is a tautology because regardless of the truth values of the statements p and q, the truth value of $\sim(p \vee q) \leftrightarrow \sim p \wedge \sim q$ is **ALWAYS TRUE**.

Also, the statements $\sim(p \vee q)$ and $\sim p \wedge \sim q$ are logically equivalent.

1. Complete the truth table below to determine if $[(p \rightarrow q) \wedge (q \rightarrow p)] \leftrightarrow (p \leftrightarrow q)$ is a tautology, a contradiction or a contingent statement.

p	q	$p \rightarrow q$	$q \rightarrow p$	$(p \rightarrow q) \wedge (q \rightarrow p)$	$p \leftrightarrow q$	$[(p \rightarrow q) \wedge (q \rightarrow p)] \leftrightarrow (p \leftrightarrow q)$
T	T					
T	F					
F	T					
F	F					

Chapter 1

LOGIC
Completing Truth Tables and Determining Tautologies (continued)

2. If two compound statements called P and Q are logically equivalent, then which of the following is true?

   (1) $P \rightarrow Q$ is a contingency.
   (2) $P \leftrightarrow Q$ is a contingency.
   (3) $P \rightarrow Q$ is a contradiction.
   (4) $P \leftrightarrow Q$ is a tautology.

   _____

4. Which of the following statements is a contradiction?

   (1) $p \wedge \sim p$ (3) $p \rightarrow \sim p$
   (2) $p \vee \sim p$ (4) $\sim p \rightarrow p$

   _____

3. Which of the following statements is a tautology?

   (1) $p \wedge \sim p$ (3) $p \rightarrow \sim p$
   (2) $p \vee \sim p$ (4) $p \leftrightarrow \sim p$

   _____

5. If $p \rightarrow q$ is a true statement and $\sim q$ is a true statement, which of the following must also be a true statement?

   (1) p (3) $\sim p$
   (2) q (4) $p \vee q$

   _____

6. Complete the truth table below to determine if $(p \rightarrow q) \leftrightarrow (\sim p \vee q)$ is a tautology, a contradiction or a contingent statement.

p	q	$p \rightarrow q$	$\sim p$	$\sim p \vee q$	$(p \rightarrow q) \leftrightarrow (\sim p \vee q)$

7. Complete the truth table below to determine if $\sim (p \rightarrow q) \leftrightarrow (p \wedge \sim q)$ is a tautology, a contradiction or a contingent statement.

p	q	$p \rightarrow q$	$\sim (p \rightarrow q)$	$\sim q$	$p \wedge \sim q$	$\sim (p \rightarrow q) \leftrightarrow (p \wedge \sim q)$

8. Since $p \rightarrow q$ is logically equivalent to $\sim p \vee q$, write a disjunction statement that is equivalent to the following conditional statement:

   "If a horse is in Oz, then the horse can change colors."

9. Since $\sim (p \rightarrow q)$ is logically equivalent to $p \wedge \sim q$, write a conjunction statement that is equivalent to the following conditional statement:

   "It is not true that if Lisa calls in sick then she will get a promotion."

Chapter 2

GEOMETRIC PROOFS (INFORMAL & FORMAL)

Congruence of Triangles, SSS, SAS, ASA, HYP-LEG

> **REMEMBER**
>
> Two triangles are congruent if their corresponding angles and corresponding sides are congruent. It is not necessary to establish that all corresponding parts of two triangles are congruent before concluding that the triangles themselves are congruent.
>
> Listed below in abbreviated form, are the key triangle congruence theorems with which you should already be familiar.
>
> (1) s.s.s. ≅ s.s.s
>
>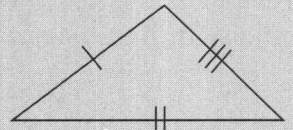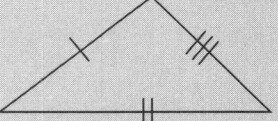
>
> (2) s.a.s. ≅ s.a.s
>
>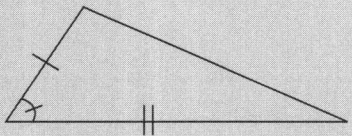
>
> (3) a.s.a. ≅ a.s.a
>
>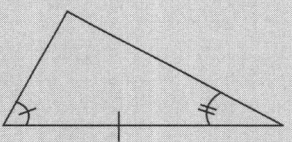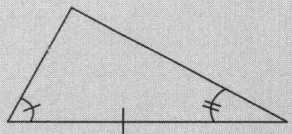
>
> (4) a.a.s. ≅ a.a.s
>
>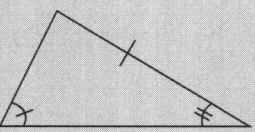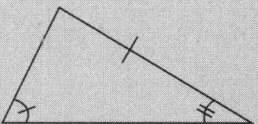
>
> (5) hypotenuse - leg theorem ($h.\ell.$ theorem) Two right triangles are congruent if the hypotenuse and leg of one are congruent to the hypotenuse and leg of the other.
>
>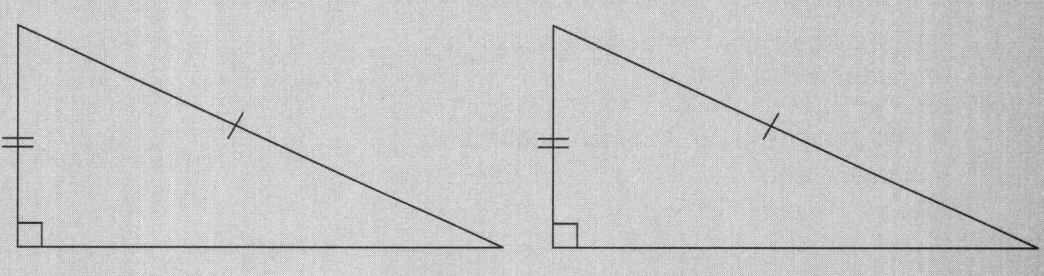

Chapter 2

GEOMETRIC PROOFS

Congruence of Triangles (continued)

REMEMBER

If two triangles are congruent, their corresponding parts are congruent. (CPCTC)

Example: In $\triangle ABC$ and $\triangle DEC$, $\angle A \cong \angle D$, $\angle B \cong \angle E$ and $\overline{AB} \cong \overline{DE}$.
If AC = 16, find the measure of DF.

Solution:

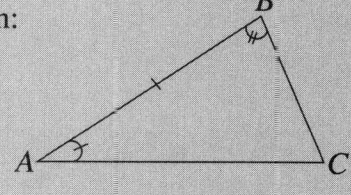 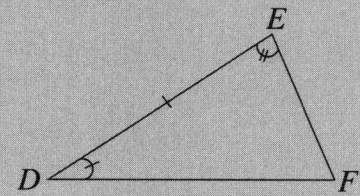

$\triangle ABC \cong \triangle DEF$ a.s.a. = a.s.a.
$\overline{AC} \cong \overline{DF}$ cpctc
$\therefore$ DF = 16

---

1. Which of the following is *not* a triangle congruence theorem?
   (1) s.s.s. ≅ s.s.s.
   (2) s.a.s. ≅ s.a.s.
   (3) s.s.a. ≅ s.s.a.
   (4) a.a.s. ≅ a.a.s.

2. $\triangle MNO \cong \triangle PQD$ with $\overline{MN} \cong \overline{PQ}$, $\overline{NO} \cong \overline{QD}$ and $\overline{MO} \cong \overline{PD}$.
   If the sum of the measures of $\angle M$ and $\angle N = 95°$, what is the degree measure of $\angle D$?

3. $\triangle ABC \cong \triangle DEF$ and the perimeter of $\triangle ABC$ is 26 inches. If the sum of the measures of two sides of $\triangle DEF$ is 19 inches, the shortest side of $\triangle DEF$ must measure:
   (1) 7 inches
   (2) 9 inches
   (3) 10 inches
   (4) cannot be determined

Chapter 2

GEOMETRIC PROOFS

Congruence of Triangles (continued)

4. In the following diagram, $\overline{AEB}$ and $\overline{CED}$ bisect each other at point E. The key theorem in proving $\triangle AEC \cong \triangle BED$ would be:

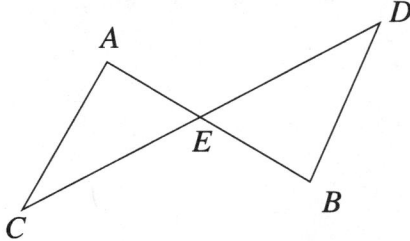

(1) a.a.s. ≅ a.a.s. (3) s.s.s. ≅ s.s.s.
(2) s.a.s. ≅ s.a.s. (4) a.a.s. ≅ a.a.s.

_____

5. $\triangle ABC \cong \triangle DEF$ with $\angle A \cong \angle D$ and $\angle B \cong \angle E$. If BC = 8, which side of $\triangle DEF$ has the same measure?
     (1) ED   (3) DF
     (2) EF   (4) None of above

_____

6. If the two legs of a right triangle measure 5 and 12 respectively, and the leg and hypotenuse of another right triangle measure 12 and 13 respectively, explain why the triangles would be congruent.

7. As pictured below, isosceles $\triangle ABC$ has median $\overline{AM}$ drawn from vertex $\angle A$. Describe briefly how each of the key congruence theorems can be used to prove that $\triangle AMB$ is congruent to $\triangle AMC$.

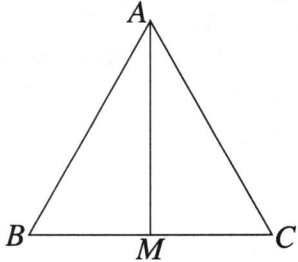

Chapter 2

GEOMETRIC PROOFS

The Parallelogram

REMEMBER

A parallelogram is a quadrilateral with opposite sides parallel. It has many special properties.

If you are given parallelogram ABCD then:

Property	Meaning
(1) opposite sides are parallel	(1) $\overline{AB} \parallel \overline{DC}, \overline{AD} \parallel \overline{BC}$
(2) opposite sides are congruent	(2) $\overline{AB} \cong \overline{DC}, \overline{AD} \cong \overline{BC}$
(3) opposite angles are congruent	(3) $\angle A \cong \angle C, \angle D \cong \angle B$
(4) consecutive angles are supplementary	(4) $m\angle A + m\angle B = 180°$
	$m\angle B + m\angle C = 180°$
	$m\angle C + m\angle D = 180°$
	$m\angle D + m\angle A = 180°$

If the diagonals are drawn then:

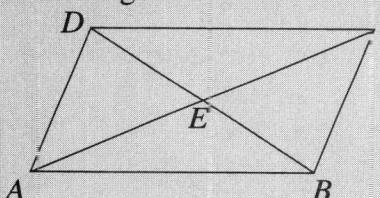

(5) the diagonals bisect each other   (5) $\overline{AB} \cong \overline{EC}, \overline{DE} \cong \overline{EB}$

---

1. In parallelogram ABCD, m $\angle$ B = 60°. Find m $\angle$ C, in degrees.

   _____

2. In parallelogram PQRS, the ratio of measure of $\angle$ Q to the measure of $\angle$ R is 1:5. Find m $\angle$ Q.

   _____

3. In parallelogram CDEF, CD = (5x - 6) and FE = (3x + 8). Find the value of x.

   _____

4. In parallelogram ABCD, m $\angle$ D = (6x + 40)° and m $\angle$ B = (4x + 70)°. Find the value of x.

   _____

5. Which statement is *not* true for every given parallelogram PQRS?

   (1) $\overline{PQ} \cong \overline{SR}$ (3) $\overline{PR} \perp \overline{SQ}$
   (2) $\angle P \cong \angle R$ (4) m $\angle$ P + m $\angle$ S = 180°

   _____

6. In parallelogram QRST, diagonals $\overline{QS}$ and $\overline{RT}$ intersect at point E. If QE = 4x + 3 and ES = 23, find the value of x.

   _____

7. In parallelogram ABCD, the measures of angles A and B are in the ratio of 1:8. Find m $\angle$ B.

   _____

19

Chapter 2

GEOMETRIC PROOFS

The Parallelogram (continued)

8. The measure of two opposite angles of a parallelogram are represented as $(5x + 40)°$ and $(3x + 50)°$. Find x

_____

9. Which statement is always true?

   (1) The diagonals of a parallelogram are congruent.
   (2) The diagonals of a parallelogram bisect the angles of the parallelogram.
   (3) The diagonals of a parallelogram bisect each other.
   (4) The diagonals of a parallelogram are perpendicular to each other.

_____

10. In parallelogram ABCD, diagonals $\overline{AC}$ and $\overline{BD}$ intersect at point E. Which statement is always true?

   (1) $\overline{AC} \perp \overline{DB}$ (3) $\triangle ABD \cong \triangle AED$

   (2) $\triangle DEC \cong \triangle AEB$ (4) $\triangle BEC \cong \triangle DEC$

_____

11. In the accompanying diagram, side $\overline{AB}$ of parallelogram ABCD is extended to E. If m $\angle$ CBE = 35°, find m $\angle$ D.

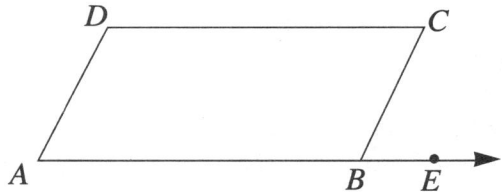

_____

12. In the accompanying diagram of parallelogram ABCD, $\overline{DF}$ is perpendicular to diagonal $\overline{AC}$ at point F. If m $\angle$ CAB = 34°, find m $\angle$ CDF in degrees.

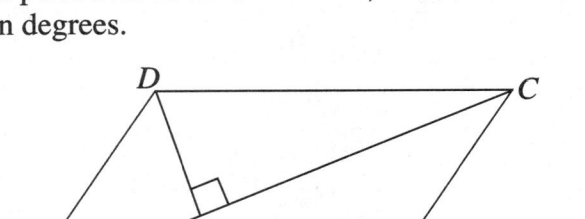

_____

13. In parallelogram ABCD, m $\angle$ A = $(3x = 40)°$ and m $\angle$ C = $(7x - 100)°$. Find the measure in degrees of $\angle$ D.

_____

14. In parallelogram ABCD, diagonals $\overline{AC}$ is drawn. If m $\angle$ D = 110° and m $\angle$ CAD = 50°, find the m $\angle$ CAB in degrees.

_____

15. In parallelogram PQRS, diagonal $\overline{PR}$ is drawn. If m $\angle$ S = 100° and m $\angle$ SRP = 30°, then which of the following statements must be true?

   (1) m $\angle$ QRP = 30°
   (2) m $\angle$ PQR = 130°
   (3) m $\angle$ QPR = 50°
   (4) m $\angle$ QPR = 30°

_____

Chapter 2

GEOMETRIC PROOFS

The Rectangle

> **REMEMBER**
>
> A rectangle is a parallelogram with a right angle. In addition to having all the properties of a parallelogram, the rectangle has several other properties.
>
> If you are given rectangle ABCD then:
>
>
>
> | Property | Meaning |
> |---|---|
> | (1) all the properties of the parallelogram are true | (1) see page 19 |
> | (2) all the angles are right and are congruent | (2) m∠A = m∠B = 90° |
> | (3) the diagonals are congruent | (3) m∠C = m∠D = 90° $\overline{AC} \cong \overline{DB}$ |

1. In rectangle PQRS with diagonals $\overline{PR}$ and $\overline{SQ}$, if PR = (4x - 10) and SQ = (7x - 40), find x.

   _____

2. In rectangle ABCD, AB = 4 and BC = 3. AC must be

   (1) 5 (3) 25
   (2) 7 (4) 4

   _____

3. In rectangle ABCD diagonal $\overline{AC}$ is drawn. If m∠DCA = 30°, find m∠CAD.

   _____

4. The diagonals of rectangle ABCD intersect at E. If DE = (3x + 1) and EB = (2x + 7), find the value of x.

   _____

5. If the measure if one angle of a parallelogram is 90°, what is the probability that the parallelogram is a rectangle?

   _____

6. In rectangle ABCD, AD = 6 and AB = 8. What is the measure of diagonal $\overline{AC}$?

   _____

7. The diagonals of rectangle ABCD intersect at E. If DE = (2x + 3) and AE = (x + 6), find the value of x.

   _____

8. The diagonals of rectangle RSTV intersect at Q. If VQ = (3x - 3) and RT = (5x - 1), find the value of x.

   _____

9. The diagonals of a rectangle must always bisect each other. (True or False).

   _____

10. In rectangle ABCD, BC = 30 and AB = 40. If diagonals $\overline{AC}$ and $\overline{BD}$ intersect at E, find the measure of $\overline{BE}$.

    _____

Chapter 2

GEOMETRIC PROOFS

The Rhombus

> **REMEMBER**
>
> A rhombus is a parallelogram with adjacent sides equal. In addition to having all the properties of a parallelogram, the rhombus has several other properties.
>
> If you are given a rhombus ABCD then:
>
>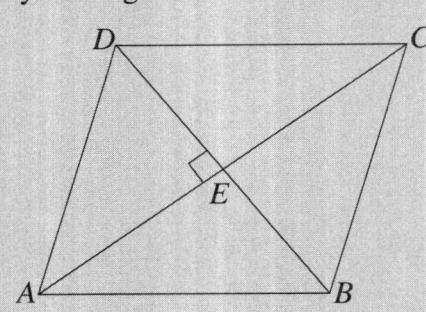
>
> | Property | Meaning |
> |---|---|
> | (1) all the properties of the parallelogram are true | (1) see page 19 |
> | (2) all sides are congruent | (2) $\overline{AB} \cong \overline{BC} \cong \overline{DC} \cong \overline{AD}$ |
> | (3) the diagonals bisect the opposite angles | (3) m∠DAC = m∠BAC
m∠DCA = m∠BCA
m∠ADB = m∠CDB
m∠ABD = m∠CBD |
> | (4) the diagonals are perpendicular to each other | (4) $\overline{DB} \perp \overline{AC}$ at E |

1. In rhombus ABCD, AB = (6x - 3) and BC = (4x + 7). Find x.

   _____

2. The lengths of the diagonals of a rhombus are 6 and 8. What is the length of a side of a rhombus?

   _____

3. In the accompanying diagram of rhombus CDEF, diagonal $\overline{FD}$ is drawn and m∠E = 40°. Find m∠CFD

   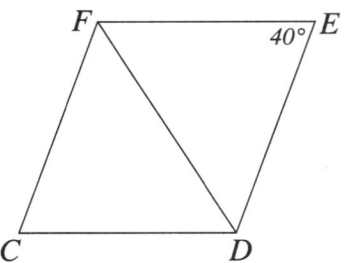

   _____

4. In rhombus ABCD, diagonal $\overline{DB}$ is congruent to side $\overline{AD}$. What is the measure of ∠A?

   _____

5. If the lengths of the diagonals of a rhombus are 10 and 24, find the length of one side of the rhombus.

   _____

6. In the accompanying diagram of rhombus ABCD, m∠ABD = 50°. Find m∠A.

   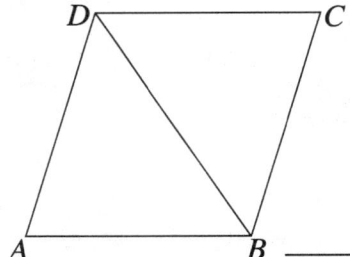

7. In rhombus ABCD, diagonal AC = 30 and side BC = 17. What is the measure of diagonal BD?

   _____

Chapter 2
GEOMETRIC PROOFS

The Square

REMEMBER

A square is a rhombus with four right angles or is a rectangle with four equal sides. Since it is both a rectangle and a rhombus, all the properties listed on pages 21 and 22 hold true for the square.

Example: If the side of a square ABCD is 5, find the diagonal $\overline{BD}$.

Solution:

$AB = DA = 5$ sides are congruent
$x^2 = 5^2 + 5^2$ pythagorean theorem
$x^2 = 25 + 25$
$x^2 = 50$
$x = \sqrt{50} = \sqrt{25 \cdot 2} = 5\sqrt{2}$ answer

1. In the accompanying diagram, quadrilateral ABCD is a square with diagonal $\overline{DB}$. Which of the statements are *not* true?

   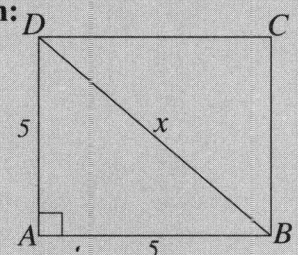

   (1) $\overline{AB} \cong \overline{CB}$
   (2) $\overline{AD} \cong \overline{CB}$
   (3) $\overline{DA} \perp \overline{AB}$
   (4) $\overline{AD} \cong \overline{DB}$

2. Which statement is false?

   (1) a square is a rectangle
   (2) a square is a rhombus
   (3) a rhombus is a square
   (4) a square is a parallelogram

3. In the accompanying diagram, PQRS is a square with diagonal $\overline{SQ}$. Which statement is *not* true?

   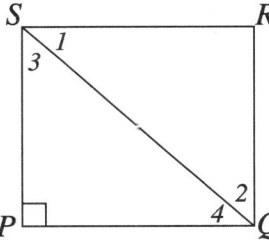

   (1) $\angle 1 \cong \angle 2$
   (2) $\angle 2 \cong \angle 3$
   (3) $\angle 4 \cong \angle P$
   (4) $\angle P \cong \angle R$

4. If the side of a square is 4, find the length of a diagonal.

   _____

5. In the accompanying diagram CDEF is a square with diagonal $\overline{CE}$ drawn. Which statement is *not* true?

   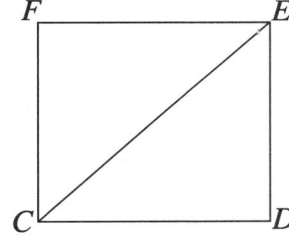

   (1) $\triangle$ CFE is isosceles
   (2) $\triangle$ CFE is a right triangle
   (3) $\triangle$ CFE $\cong$ $\triangle$ CDE
   (4) $\triangle$ CDE is equilateral

6. In square ABCD diagonal $\overline{AC}$ is drawn. How many degrees are there in the measure of $\angle$ ACB?

   _____

7. Find the diagonal of a square whose perimeter is 28.

   _____

Chapter 2

GEOMETRIC PROOFS

The Trapezoid and Isosceles Trapezoid

> **REMEMBER**
>
> A trapezoid is a quadrilateral that has two and only two sides parallel. The parallel sides are called the bases and the non-parallel sides are called the legs. An isosceles trapezoid is a trapezoid which has congruent legs.
>
> **Example:** Given isosceles trapezoid ABCD with $\overline{AB} \parallel \overline{CD}$, AB = 4, CD = 14, and AD = 13. Find the length of an altitude of trapezoid ABCD.
>
> **Solution:** Draw in altitudes $\overline{AE}$ and $\overline{BF}$ such that rectangle ABFE is formed and EF = 4.
> Since ABCD is isosceles, $\triangle$ ADE can be proven to be congruent to $\triangle$ BFC and $\overline{DE} \cong \overline{FC}$. ∴ DE = FC = 5.
> By the pythagorean theorem $x^2 + 5^2 = 13^2$
> $x^2 + 25 = 169$
> $x^2 = 144$
> x = 12 answer

1. In the accompanying diagram, isosceles trapezoid CDEF has bases of lengths 8 and 18 and an altitude of length = to 12. Find CD.

   _____

2. In the accompanying diagram of trapezoid ABCD, CD = 8, m $\angle$ A = 30° m $\angle$ D = 90. Find the length of leg AB.

   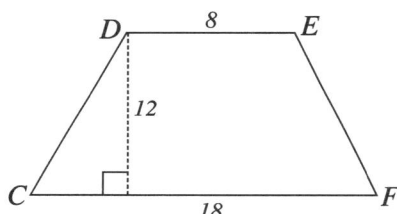

   _____

3. In isosceles trapezoid ABCD, $\overline{AB} \parallel \overline{CD}$, AB = 18, CD = 6 and AD = 10. Find the length of an altitude of ABCD.

   _____

4. In the accompanying diagram, isosceles trapezoid ABCD has bases $\overline{AB}$ and $\overline{DC}$, and diagonals $\overline{AC}$ and $\overline{BD}$ are drawn. Which statement is *not* true?

   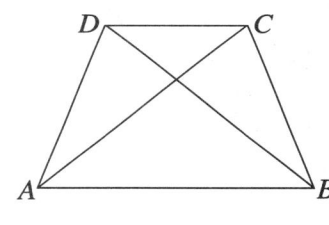

   (1) $\overline{AD} \cong \overline{BC}$
   (2) $\overline{AC} \cong \overline{BD}$
   (3) $\overline{AB} \cong \overline{DC}$
   (4) $\overline{AB} \parallel \overline{DC}$

   _____

5. ABCD is an isosceles trapezoid with bases $\overline{AB}$ and $\overline{DC}$. If AD = 3x + 4 and BC = x + 12, find x.

   _____

6. CDEF is a trapezoid with $\overline{CD} \parallel \overline{FE}$. If m $\angle$ F and m $\angle$ C are in the ratio 1:4, find the measure of $\angle$ F.

   _____

24

Chapter 2

GEOMETRIC PROOFS

Formal Proofs (hypothesis to conclusion)

REMEMBER

Example:

Given: △ABC, $\overline{AEB}$, $\overline{AFC}$, D is the midpoint of $\overline{BC}$, $\overline{ED} \cong \overline{FD}$, ∠EDC ≅ ∠FDB.

Prove: △ABC is isosceles.

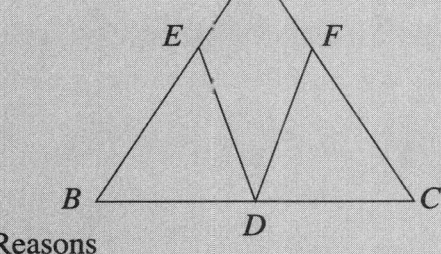

Proof:

Statements	Reasons
1. △ABC, $\overline{AEB}$, $\overline{AFC}$, D is the midpoint of BC.	1. Given.
2. $\overline{BD} \cong \overline{DC}$.	2. Definition of a midpoint.
3. $\overline{ED} \cong \overline{FD}$.	3. Given.
4. ∠EDF ≅ ∠EDF.	4. Reflexive property of congruence.
5. ∠EDC ≅ ∠FDB.	5. Given.
6. ∠EDC − ∠EDF ≅ ∠FDB − ∠EDF or ∠EDB ≅ ∠FDC.	6. Subtraction postulate of congruent angles.
7. △EDB ≅ △FDC.	7. s.a.s. ≅ s.a.s. (steps 2, 3, and 6).
8. ∠B ≅ ∠C.	8. Corresponding angles of congruent triangles are congruent.
9. $\overline{AB} \cong \overline{AC}$.	9. If two angles of a triangle are congruent then the sides opposite these angles are congruent.
10. △ABC is isosceles.	10. Definition of an isosceles triangle.

---

1. **Given:** quadrilateral ABCD with point E on $\overline{AC}$, $\overline{AB} \cong \overline{AD}$, $\overline{AC}$ biscents ∠DAB, $\overline{BE}$ and $\overline{DE}$ are drawn.

   Prove: ∠BEC ≅ ∠DEC

   Next to each numeral, give a reason for each statement in the proof.

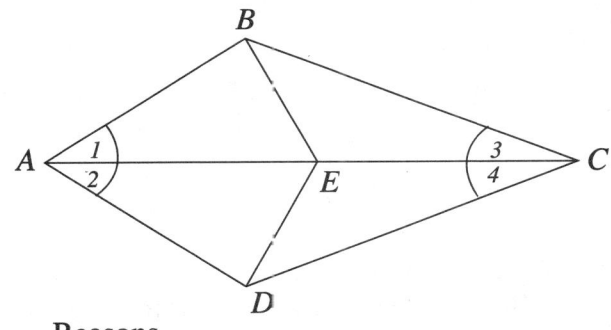

Proof:

Statements	Reasons
1. Quadrilateral ABCD with point E on $\overline{AC}$, $\overline{AB} \cong \overline{AD}$, $\overline{AC}$ bisects ∠DAB, $\overline{BE}$ and $\overline{DE}$ are drawn.	1.
2. ∠1 ≅ ∠2	2.
3. $\overline{AC} \cong \overline{AC}$	3.
4. △ABC ≅ △ADC	4.
5. $\overline{BC} \cong \overline{CD}$, ∠3 ≅ ∠4	5.
6. $\overline{EC} \cong \overline{EC}$	6.
7. △BEC ≅ △DEC	7.
8. ∠BEC ≅ ∠DEC	8.

25

Chapter 2

GEOMETRIC PROOFS

Formal Proofs (continued)

2. Given: quadrilateral ABCD, $\overline{BFE}$, $\overline{CFD}$, $\overline{ADE}$, $\overline{BE}$ bisects $\overline{CD}$, $\overline{AE}$ ∥ $\overline{BC}$.

   Prove: $\overline{BF} \cong \overline{FE}$.

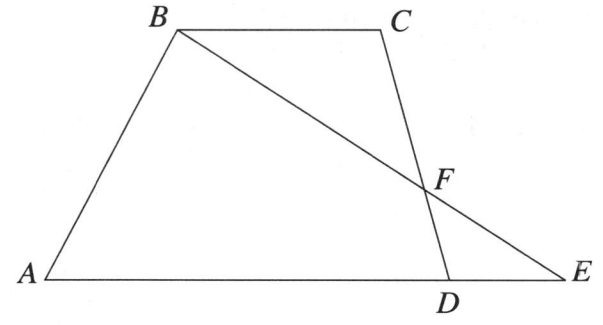

3. Given: △ ABC, $\overline{BEA} \cong \overline{BDC}$, $\overline{AD}$ and $\overline{CE}$ intersect at F, and ∠ FAC ≅ ∠ FCA.

   Prove: $\overline{FE} \cong \overline{FD}$.

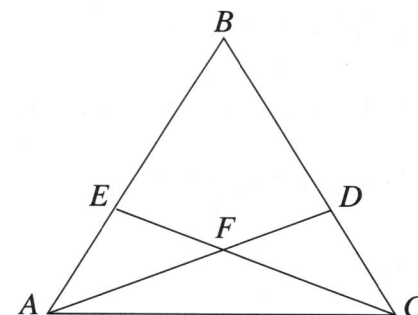

26

Chapter 2

GEOMETRIC PROOFS

Formal Proofs (continued)

4. Given: parallelogram ABCD with diagonal $\overline{AEFC}$, $\overline{AE} \cong \overline{FC}$.

   Prove: $\overline{BF} \parallel \overline{DE}$.

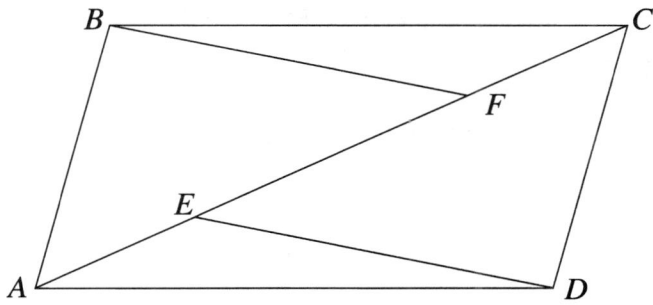

5. Given: quadrilateral ABCD, with diagonals $\overline{AC}$ and $\overline{BD}$ intersecting at point E, $\angle EAD \cong \angle ECB$, $\overline{BD}$ bisects $\overline{AC}$.

   Prove: ABCD is a parallelogram

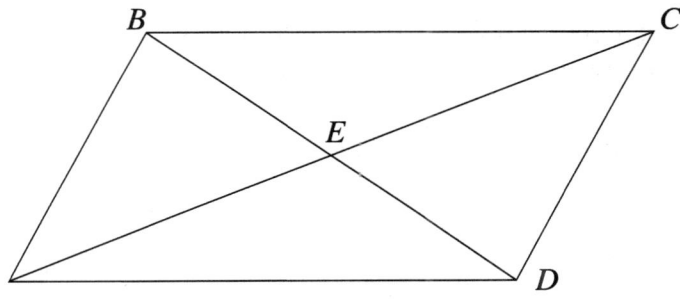

Chapter 2
GEOMETRIC PROOFS

Formal Proofs (continued)

6. Given: parallelogram ABCD, M is the midpoint of $\overline{AD}$, $\overline{BM} \cong \overline{MC}$.

    Prove: ABCD is a rectangle.

    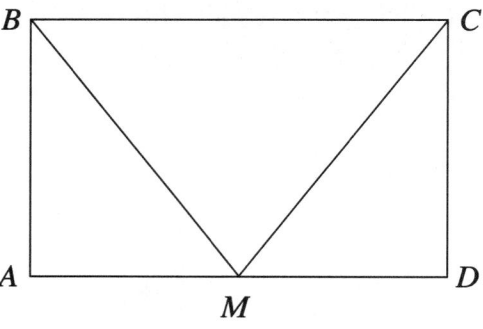

7. Given: rectangle ABCD, $\overline{AFED}$, $\overline{AF} \cong \overline{ED}$.

    Prove: $\angle EBA \cong \angle FCD$.

    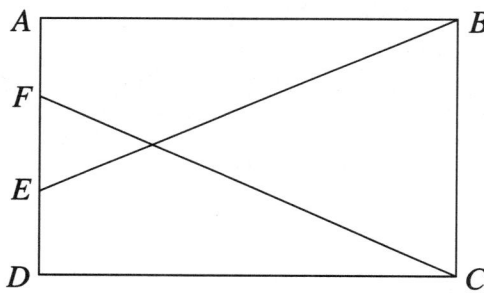

Chapter 2
GEOMETRIC PROOFS

Formal Proofs (continued)

8. Given: quadrilateral ABCD with diagonals $\overline{AEFC}$, $\overline{DE} \perp \overline{AC}$, $\overline{BF} \perp \overline{AC}$, $\angle EDC \cong \angle ABF$, $\overline{DE} \cong \overline{BF}$.

   Prove: ABCD is a parallelogram.

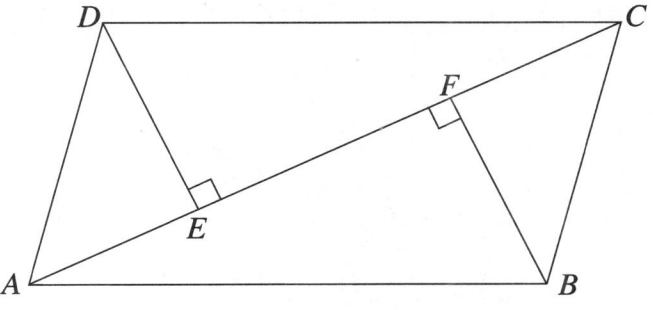

9. Given: $\triangle ABC$, altitudes $\overline{BD}$ and $\overline{CE}$ are drawn and $\overline{BE} \cong \overline{CD}$

   Prove: $\overline{BD} \cong \overline{CE}$

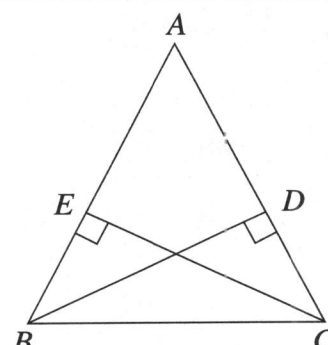

Chapter 2

GEOMETRIC PROOFS

Similarity of Triangles

REMEMBER

Two similarity theorems to remember are:
(1) If two angles of one triangle are congruent to two angles of another triangle, the triangles are similar. (aa ≅ aa)

(2) If two triangles are similar, then their corresponding sides are in proportion.

Example: Given: Right △ ABC with right angle C.
$\overline{BE} \perp \overline{AB}$.

Prove: $\dfrac{AB}{AC} = \dfrac{AE}{AD}$

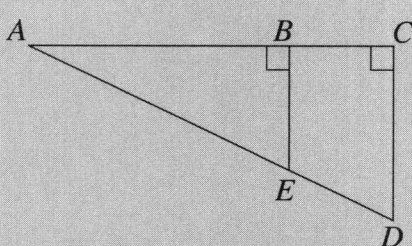

Proof:

Statements	Reasons
1. Right △ ABC with rt. ∠ C and $\overline{BE} \perp \overline{AB}$.	1. Given
2. ∠ ABC is a right angle.	2. Perpendicular lines form right angles.
3. ∠ ABE ≅ ∠ C	3. All right angles are congruent.
4. ∠ A ≅ ∠ A	4. Reflexive property of congruence.
5. △ ABE ~ △ ACD	5. aa ≅ aa similarity theorem
6. $\dfrac{AB}{AC} = \dfrac{AE}{AD}$	6. Corresponding sides of similar triangles are in proportion.

1. Which triangles are always similar?

　　(1) equilateral triangle　　　　　　(3) right triangles

　　(2) isosceles triangles　　　　　　(4) scalene triangles

30

Chapter 2
GEOMETRIC PROOFS

Similarity Formal Proofs

2. Given: △ ABC, altitudes $\overline{BE}$ and $\overline{AD}$ are drawn.

   Prove: $\dfrac{AC}{BC} = \dfrac{AD}{BE}$

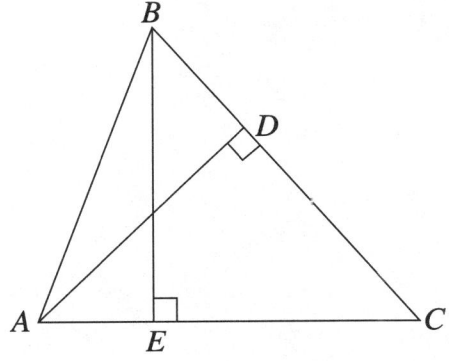

3. Given: ABCD is a parallelogram, $\overline{ADE}$, $\overline{BFE}$ are drawn.

   Prove: (a) $\dfrac{BC}{DE} = \dfrac{BF}{FE}$

   (b) $\dfrac{AD}{DE} = \dfrac{BF}{FE}$

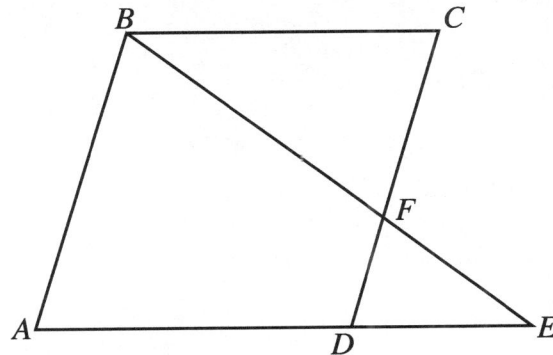

Chapter 2
GEOMETRIC PROOFS

Proportions in Similar Triangles

> **REMEMBER**
>
> In a proportion, the product of the means equals the product of the extremes.
>
> **Example:** In the accompanying diagrams of $\triangle ABC$ and $\triangle DEF$, $\angle A \cong \angle D$ and $\angle B \cong \angle E$. If $AC = 8$, $DF = 6$ and $AB = 4$, find DE.
>
>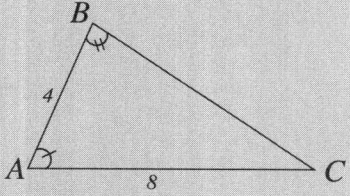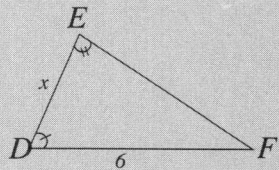
>
> **Solution:**
>
> $\triangle ABC \sim \triangle DEF$ aa $\cong$ aa
>
> $\dfrac{AB}{DE} = \dfrac{AC}{DF}$ Corresponding sides of similar triangles are in proportion.
>
> $AC \cdot DE = DF \cdot AB$ Product of means equals product of extremes.
>
> $8x = 24$ Substitution
>
> $x = 3$ ans.

1. In $\triangle ABC$, $\overline{DE} \parallel \overline{AC}$, $DE = 3$ and $AC = 8$. If $BD = 6$, find the measure of AB.

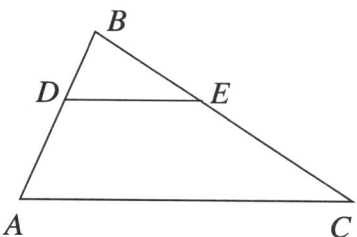

2. The sides of a triangle are 5, 7 and 10 respectively. Find the perimeter of a similar triangle whose shortest side measures 15.

32

Chapter 2
GEOMETRIC PROOFS

Proportions in Similar Triangles

3. In the accompanying diagram, triangles ABC and PQR are right triangles with ∠ C and ∠ R being right angles and ∠ A ≅ ∠ P. If AB = 12, PQ = 9 and AC = 8, find PR.

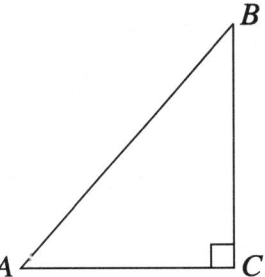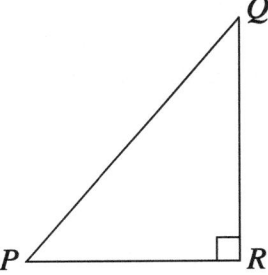

4. The sides of a triangle are 3, 7 and 9. Find the length of the shortest side of a similar triangle whose longest side is 36

    (1) 28   (3) 9
    (2) 12   (4) 6

5. In the accompanying diagram $\overline{AB} \parallel \overline{CD}$ and $\overline{AD}$ intersects $\overline{BC}$ at point E. If AB = 24, BE = 15 and CD = 8, find the measure of CE.

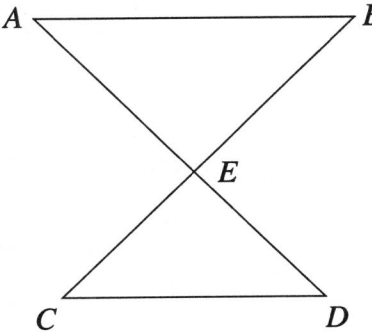

33

Chapter 2

GEOMETRIC PROOFS

Proportions in Similar Triangles

6. In the accompanying diagram of △ ABC, $\overline{DE} \parallel \overline{AC}$, BD = 8, DA = 6 and EC = 9. Find the measure of BC.

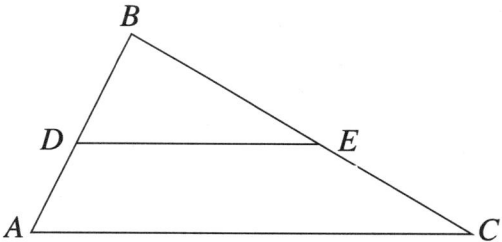

7. In the accompanying diagram $\overline{AB} \perp \overline{BE}$, $\overline{AC} \perp \overline{CD}$, AC = 20 and CD = 10. If BC = CD, find the measure of BE.

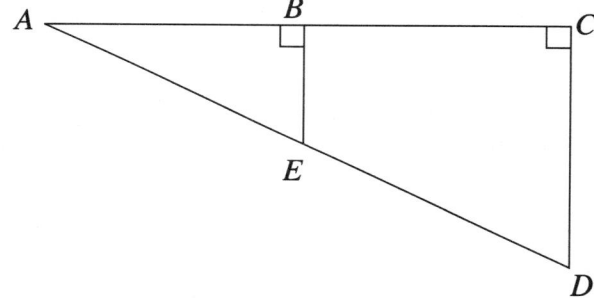

8. A 10′ ladder and a 15′ ladder are placed against the same wall and make the same number of degrees in their angles with the ground. If the 10′ ladder reaches a height of 6′ against the wall. How much farther up the wall does the 15′ ladder reach?

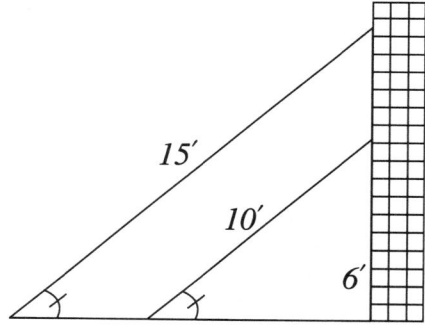

Chapter 2

GEOMETRIC PROOFS

Proportions in Similar Triangles

9. A boy scout troop was asked to approximate the distance across a marshy tract of land from point P to D by marking off right triangles PDQ and PCL with segments $\overline{CPD}$ and $\overline{LPQ}$, as shown below. If LC = 9 feet, CP = 27 feet and DQ = 81 feet, find PD in feet.

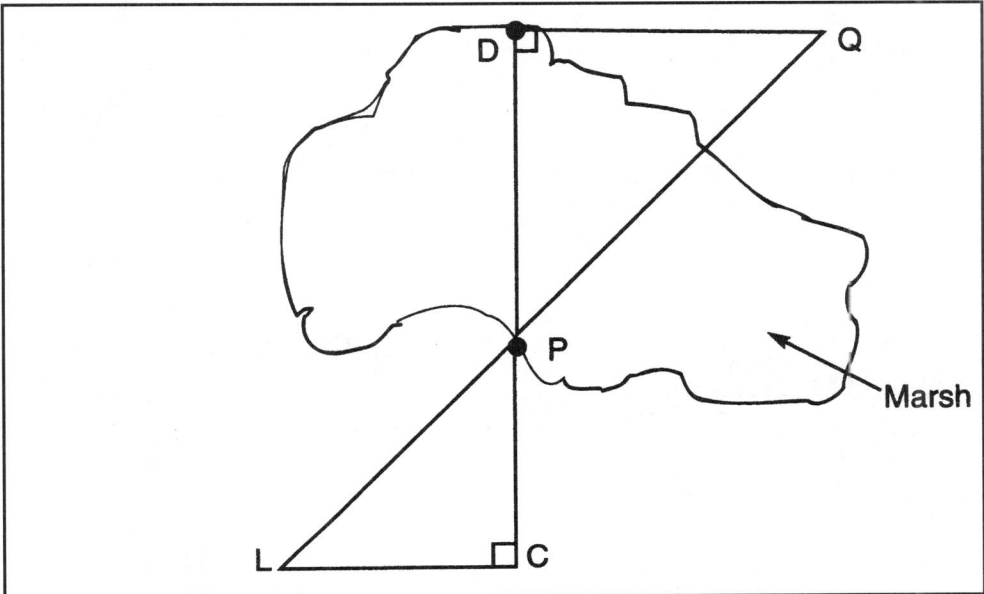

10. At the same time of day a telephone pole and a yard stick cast a shadow of 20 feet and 4 feet respectively as shown in the diagram below. If the angle of elevation of the sun is the same for both the pole and the yard stick, determine the height of the telephone pole.

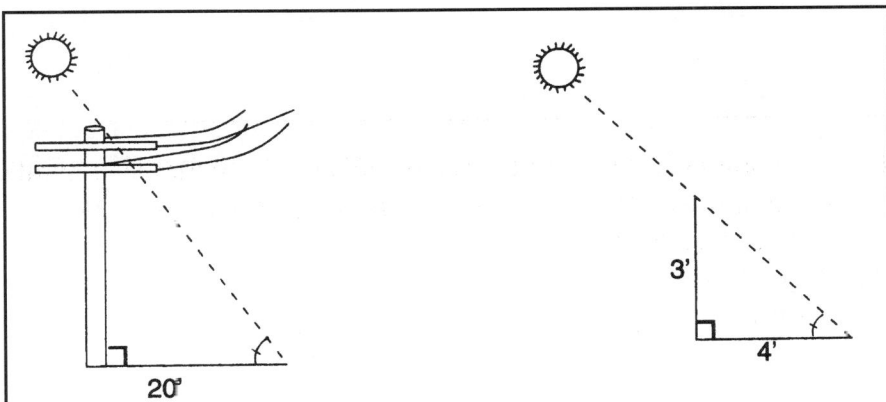

35

Chapter 3

MEASUREMENT IN POLYGONS

The Sum of the Measures of the Angles of a Triangle

REMEMBER

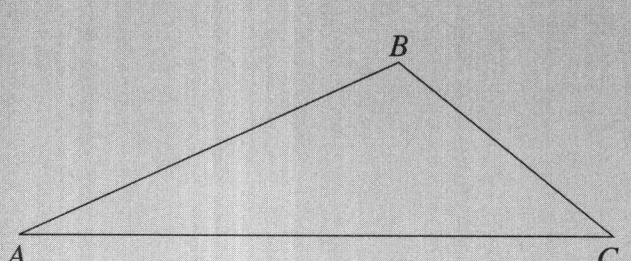

The sum of the interior angles of a triangle = 180°

Example: The measures of the angles of a triangle ABC are ∠A = x°, ∠B = 3x° and ∠C = (x + 60)°. Find the number of degrees in ∠C.

Solution: x + 3x + x + 60 = 180
5x + 60 = 180
5x = 180 − 60 = 120
x = 24, therefore **Answer:** ∠C = (x + 60)° = 24 + 60 = 84°

---

1. In triangle ABC, m∠A = x°, m∠B = (x + 10)° and m∠C = (3x + 20)°. What is the number of degrees in ∠A?

   _____

2. In triangle DEF, m∠E = (x + 10)°, m∠D = (3x + 30)° and m∠F = (5x + 50)°. How many degrees are there in ∠F?

   _____

3. What is the number of degrees in an acute angle of an isosceles right triangle?

   _____

4. If the measure of each base angle of an isosceles triangle is 20°. Explain why the vertex angle must be = 140°.

   _____

5. Three angles of a triangle are in the ratio 5:6:7. Find the number of degrees in the smallest angle.

   _____

6. Two angles of a triangle are equal in measure and the third angle is 110°. Find the number of degrees in one of the two equal angles.

   _____

7. If the measure of each base angle of an isosceles triangle is x°, then the measure of the vertex angle is represented by:
   (1) (2x − 180)°   (3) (180 − x)°
   (2) 2x°   (4) (180 − 2x)°

   _____

Chapter 3

MEASUREMENT IN POLYGONS

Exterior Angle Theorem in a Triangle

REMEMBER

The exterior angle theorem may be stated in two different ways:

a An exterior angle of a triangle is **equal** to the sum of the two non-adjacent interior angles of the triangle. $m\angle A + m\angle B = m\angle BCD$

b An exterior angle of a triangle is greater than either of the two non-adjacent interior angles of the triangle. $m\angle BCD > m\angle A$, $m\angle BCD > m\angle B$

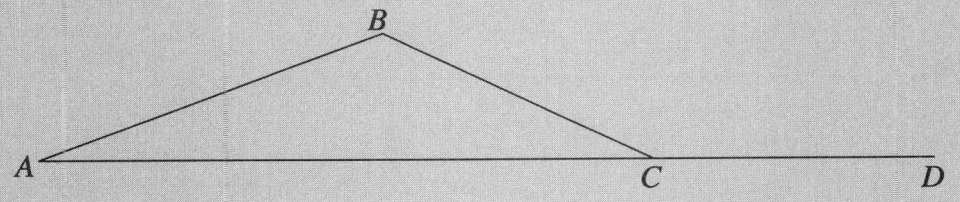

1. In an equilateral triangle, what is the measure of one of its exterior angles?

2. In right triangle ABC $\angle C = 90°$. How many degrees in $m\angle A$ if the measure of exterior $\angle ABD = 130°$?

3. In triangle ABC, $\overline{AB} = \overline{CB}$, point D is on $\overline{AB}$ and $m\angle CBD = 124°$. Find $m\angle A$.

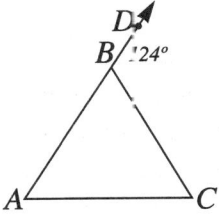

4. The measure of an exterior angle of a triangle can *NOT* be

   (1) >90° (3) = 90°

   (2) between 90° and 180° (4) >180°

5. In triangle ABC, the $m\angle BCD = 100°$ and the $m\angle BAC = 35°$. Find the $m\angle ABC$

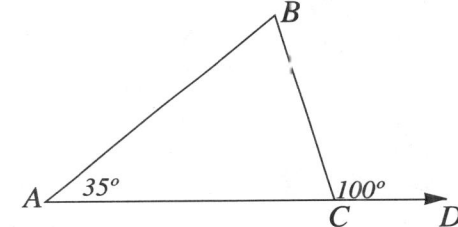

6. In isosceles triangle ABC, $\overline{AB} \cong \overline{AC}$. In the $m\angle C = (6x + 10)°$ and the $m\angle B = (3x + 40)°$, find the measure of an exterior angle at vertex A.

7. In triangle ABC, $m\angle A$ is 3 times as large as $m\angle B$. If the measure of the exterior $\angle C$ is 120°, what is the $m\angle B$?

Chapter 3

MEASUREMENT IN POLYGONS

Exterior Angle Theorem in a Triangle (continued)

8. In $\triangle$ ABC drawn below, $\overline{BD}$ is drawn such that $\overline{BD} \cong \overline{AB}$. If $\angle$ A = 50° and $\angle$ ABD = 70° more than $\angle$ DBC, find m $\angle$ C.

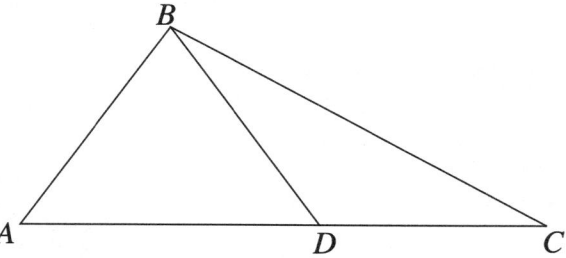

9. In $\triangle$ DEF, $\angle$ D = x° and $\angle$ E = (3x)°. Explain why exterior $\angle$ DFG must be (4x)°.

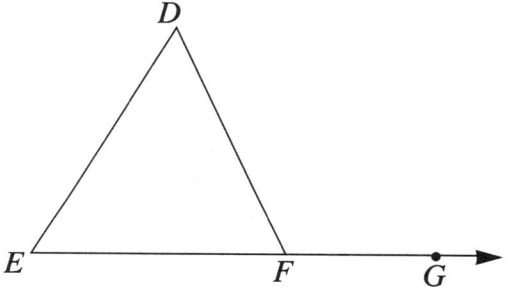

10. Use the information in the diagram below to determine the measure of $\angle$ B.

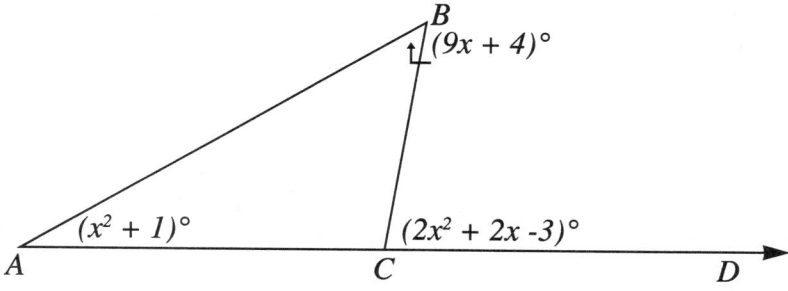

Chapter 3

MEASUREMENT IN POLYGONS

The Bisector of the Vertex Angle of Isosceles Triangle Theorems

> **REMEMBER**
>
> The bisector of the vertex angle of an isosceles triangle is perpendicular to the base.
> **the converse of this is also true**
> If a line drawn to the vertex angle of an isosceles triangle is perpendicular to the base then the line bisects the vertex angle.
>
> The line that is perpendicular to the base of the isosceles triangle is the **altitude, median** or **vertex angle bisector**.
>
>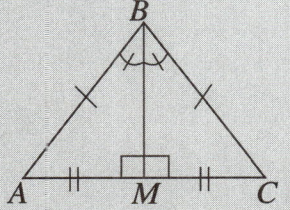

1. If the measure of vertex ∠ B of an isosceles triangle is 80° and bisector $\overline{BD}$ is drawn to the base $\overline{AC}$. Find the measure of ∠ ADB

   _____

2. In isosceles △ ABC, $\overline{AB} \cong \overline{BC}$ and $\overline{BD}$ bisects ∠ ABC. Explain why △ ABD must be a right triangle.

   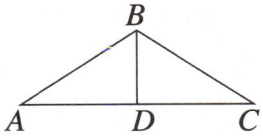

3. Each base angle of isosceles △ ABC measures 42°. If $\overline{BD}$ is an altitude then m ∠ CBD is:

   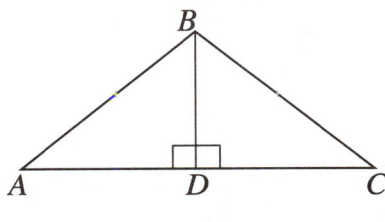

   (1) 42°   (3) 96°
   (2) 48°   (4) 90°

   _____

4. In isosceles △ RST, ∠ S is the vertex angle and $\overline{SW}$ is an altitude drawn to $\overline{RT}$. If m ∠ R = 65°, find m ∠ RSW.

   _____

5. In △ DEF, $\overline{DE} \cong \overline{EF}$, $\overline{EG}$ bisects ∠ DEF, ∠ D = (x + 5)° and ∠ DEF = (3x + 10)°. Find the measure of ∠ DEG.

   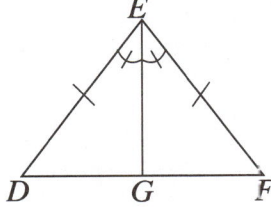

   _____

6. If each base angle of an isosceles triangle is represented by x^2 and the sum of these two base angles equal 72°, find the measure of a base angle.

   _____

7. If the angle formed by the bisector of the vertex angle of an isosceles triangle = 50°, find a base angle.

   _____

39

Chapter 3

MEASUREMENT IN POLYGONS

Inequality Theorem of Triangles

REMEMBER

In triangles, the longest side is opposite the largest angle and the shortest side is opposite the smallest angle. The largest angle is opposite the longest side and the smallest angle is opposite the shortest side.

For a triangle to exist, the sum of the two smallest sides must be greater than the largest side. If **a** and **b** are the shortest sides then $a + b > c$

Example: Given the sets of numbers representing the sides of a triangle, explain whether **a triangle can exist** with the following measures.
a {6,4,2} *b* {4,6,8} *c* {5,5,12} *d* {5,12,12}

Solution:
a **Not a triangle** because $4 + 2 \not> 6$ *b* **Is a triangle** because $4 + 6 > 8$
c **Not a triangle** because $5 + 5 \not> 12$ *d* **Is a triangle** because $5 + 12 > 12$

---

1. If two sides of a scalene triangle measure 12 and 14, the length of the third side could be:
   (1) 12 (3) 20
   (2) 2 (4) 26

2. In triangle RST, m∠R = 70° and m∠S = 30°. What is the longest side of the triangle? Explain how you arrive at your answer.

3. In any triangle ABC, which statement is always true?
   (1) AB = BC + CA (3) AB + BA > BC
   (2) AB + BC < CA (4) AB + AC > BC

4. In triangle ABC, AB = 7, AC = 4 and BC = 10. What is the smallest angle in triangle ABC?

5. In the accompanying diagram of △ABC, AC is extended to D. If m∠BCD = 120° and m∠B = 55°, which is the longest side of △ABC?

   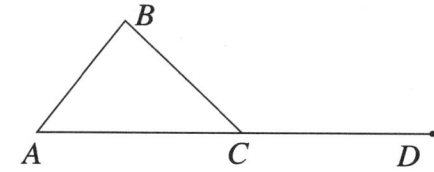

6. In triangle ABC, m∠A = 80° and AB > AC. What is the smallest angle?

40

Chapter 3

MEASUREMENT IN POLYGONS

Inequality Theorem of Triangles (continued)

7. In the △ ABC drawn below, $\overline{BD} \cong \overline{AB}$, if ∠ A = 50° and ∠ ABD is 70° more than ∠ DBC then the degree measure of ∠ C is:

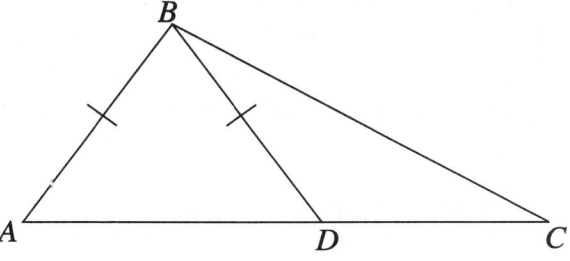

8. In a right triangle the side opposite the right angle is called the hypotenuse. Explain why the hypotenuse has to be the longest side of a right triangle.

9. The diagram below represents the relative position of three towns on a map. If town B is 40 miles northeast of town A and town C is 60 miles southeast of town B, which inequality best describes the distance between town A and town C.

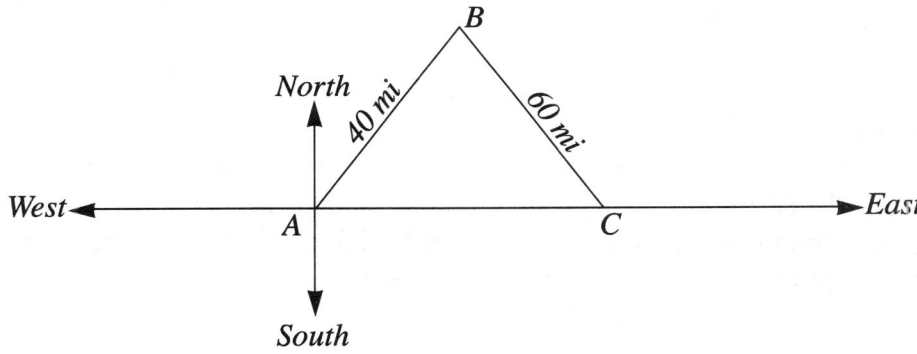

(1) AC > AB + BC

(2) AC < AB + BC

(3) AC = AB + BC

(4) AC > AB and BC > AC

10. From the results of your answer in exercise 9. What conclusion can you make concerning the sum of the measures of the two sides of a triangle with respect to the measure of the third side?

Chapter 3

MEASUREMENT IN POLYGONS

Inequality Theorem of Triangles (continued)

11. Which triple could never be the measures of three sides of a triangle?
    - (1) 2,3,4
    - (2) 3,4,5
    - (3) 3,4,7
    - (4) 9,12,14

12. In △ EFG, FE = 8 and FG = 15. Which inequality best describes the measure of EG?
    - (1) EG > 7
    - (2) EG < 23
    - (3) 7 ≤ EG ≤ 23
    - (4) 7 < EG < 23

13. If the lengths of two sides of a triangle are 5 and 11, which would be the length of the third side?
    - (1) 6
    - (2) 10
    - (3) 16
    - (4) 22

14. In △ DEF, m∠D = 70° and $\overline{DE} > \overline{DF}$. What angle is the smallest angle of △ DEF?

15. In the map below the distance from Albany to Buffalo is 240 miles and the distance from New York City to Albany is 138 miles. Explain why the distance from New York City to Buffalo cannot be less than 378 miles.

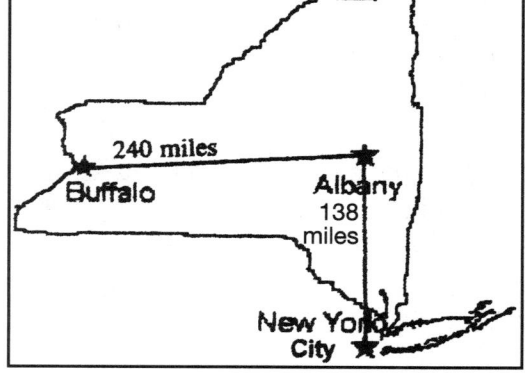

Chapter 3

MEASUREMENT IN POLYGONS

Parallel and Intersecting Lines

> **REMEMBER**
>
> Two lines cut by a transversal are parallel if:
>
> a The alternate interior angles are congruent
> b The corresponding angles are congruent
> c The **interior** or **exterior** angles on the same side of the transversal are supplementary
>
>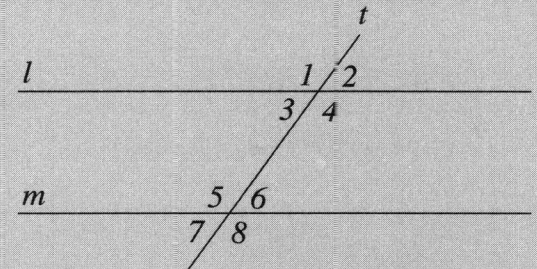
>
> If $l \parallel m$ then:
>
> a Alternate interior angles are $\cong$
> $\angle 3 = \angle 6, \angle 4 = \angle 5$
>
> b Corresponding angles are $\cong$
> $\angle 2 = \angle 6, \angle 4 = \angle 8, \angle 1 = \angle 5, \angle 3 = \angle 7$
>
> c Interior angles on the same side of the transversal are supplementary
> $\angle 3 + \angle 5 = 180°, \angle 4 + \angle 6 = 180°$
> d Exterior angles on the same side of the transversal are supplementary
> $\angle 1 + \angle 7 = 180°, \angle 2 + \angle 8 = 180°$
>
> The converse of this is true: If two lines are **parallel** the alternate interior angles are congruent, the corresponding angles are congruent, the interior angles on the same side of the transversal are supplementary and the exterior angles on the same side of the transversal are supplementary.

1. In the diagram below, m $\angle 1 = (2x + 12)°$ and m $\angle 2 = (3x + 18)°$. If $\overleftrightarrow{AB} \parallel \overleftrightarrow{CD}$, find the degree measure of $\angle 1$.

   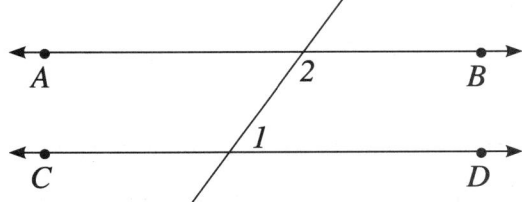

2. In the accompanying diagram, $\overleftrightarrow{AB} \parallel \overleftrightarrow{CD}$ and transversal $\overleftrightarrow{EF}$ intersects $\overleftrightarrow{AB}$ at G and $\overleftrightarrow{CD}$ at H. If $m \angle AGH = (40 - x)°$ and $m \angle GHC = (6x + 10)°$, find the value of x.

   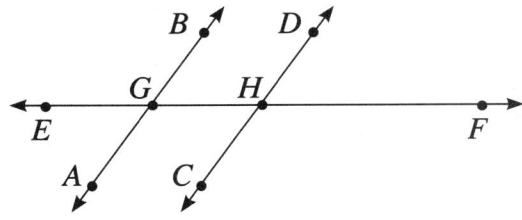

43

Chapter 3

MEASUREMENT IN POLYGONS

Parallel and Intersecting Lines (continued)

3. In the diagram below, $m\angle EGB = (6x-10)°$ and $m\angle GHD = (4x+20)°$. If $\overleftrightarrow{AB} \parallel \overleftrightarrow{CD}$ find the m $\angle CHG$.

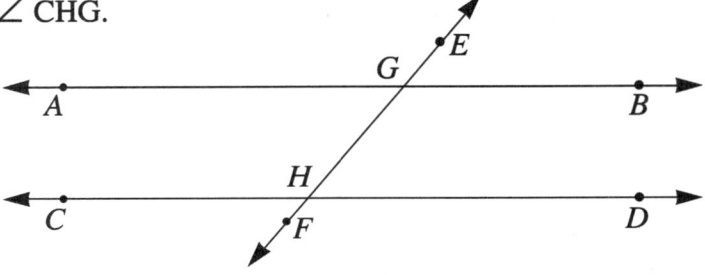

4. In the figure below $\overleftrightarrow{MN} \parallel \overleftrightarrow{QP}$ and $\overleftrightarrow{BD} \perp \overleftrightarrow{QP}$. If m $\angle CBN = 75°$, find the measure of $\angle 1$.

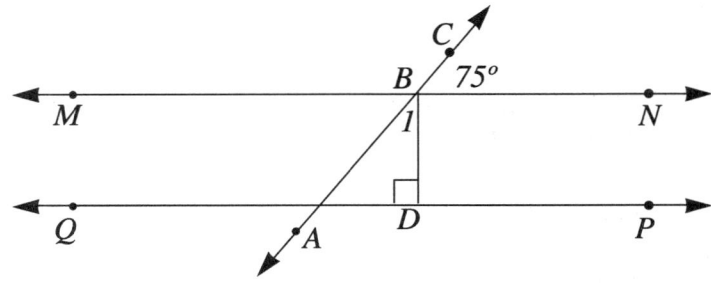

5. If $\ell \parallel m$ and t is a transversal, find the degree measure of $\angle 1$ if $\angle 1 = (5x+10)°$ and $\angle 2 = (x+110)°$.

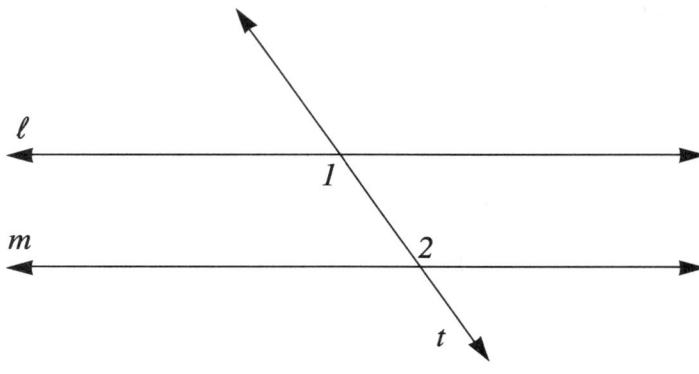

6. In the accompanying diagram, $\overleftrightarrow{AB} \parallel \overleftrightarrow{CD}$, and $\overleftrightarrow{EF}$ intersects $\overleftrightarrow{AB}$ at G and $\overleftrightarrow{CD}$ at H, $\overline{MH}$ bisects $\angle CHG$. If m $\angle AGH = (2x-30)°$ and m $\angle GHD = (x+10)°$, find m $\angle MHC$

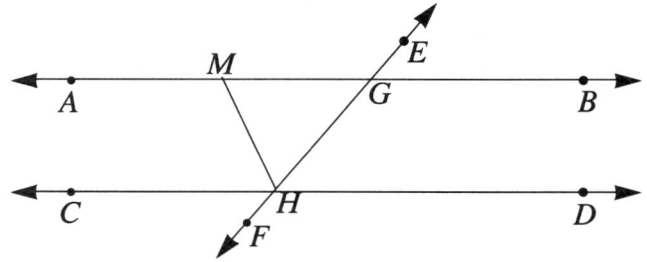

44

Chapter 3

MEASUREMENT IN POLYGONS

Parallel and Intersecting Lines (continued)

7. In the following diagram, lines ℓ and m are intersected by transversal t. If the $m\angle 1 = 40°$ and the $m\angle 2 = 130°$, are ℓ and m parallel? Explain your answer.

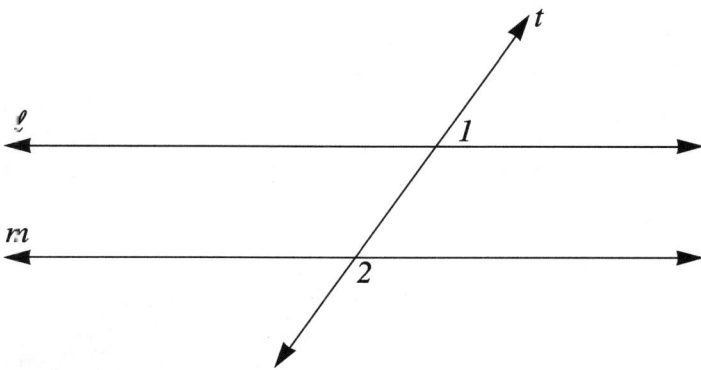

8. If $\overleftrightarrow{DE} \parallel \overline{AB}$, $m\angle DCA = 40°$ and the $m\angle ECB = 70°$, find the measures of each angle of $\triangle ABC$. Explain how you arrived at the angle measurements.

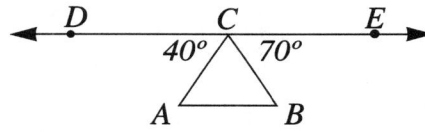

9. In the diagram below $\overleftrightarrow{AB} \parallel \overleftrightarrow{CD}$ and are intersected by two transversals, ℓ and m. Using the angle measurements denoted on the figure, show that the measures of $\angle 1$, $\angle 2$, $\angle 3$ and $\angle 4$ add up to 360°. Explain how you arrived at your answer.

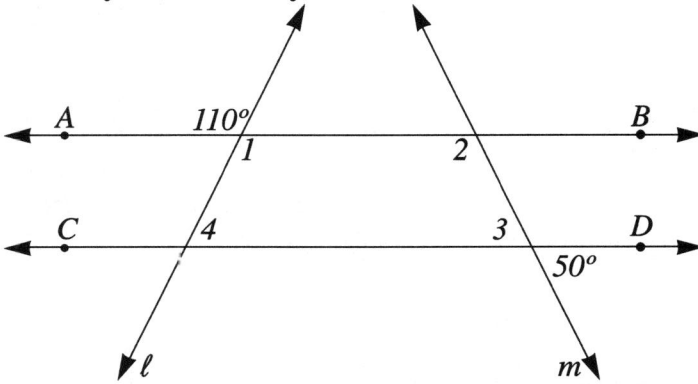

10. In the diagram below, both $\overline{AB}$ and $\overline{DC}$ are $\perp \overline{BC}$. If $m\angle DAB = (x+56)°$ and $m\angle ADC = x°$, find the value of x.

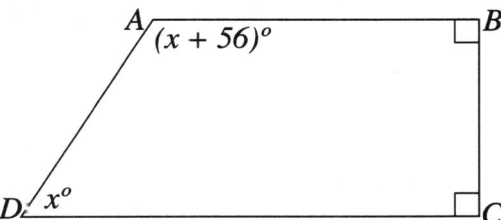

45

Chapter 3

MEASUREMENT IN POLYGONS

Sum of Interior and Exterior Angles in Polygons

> **REMEMBER**
>
> The sum of the measures of the interior angles of a polygon = $180(n - 2)°$ where n is the number of sides.
>
> The sum of the measures of the exterior angles of a polygon is always equal to 360°
>
>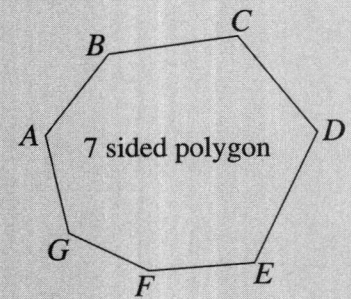
> 7 sided polygon
>
> Sum of the interior angles $= (n - 2)180°$
> $= (7 - 2)180°$
> $= 5 \cdot 180°$
> $= 900°$
>
> Sum of the exterior angles of *any* polygon is always 360°

1. What is the sum of the interior angles of an eleven sided polygon.

   _____

2. If the sum of the angles of a polygon is 1800°, how many sides in the polygon?

   _____

3. Is it possible to have a polygon whose sum of the interior angles is 500°? Explain your answer.

4. If the sum of the interior angles of a polygon is 5040°, then the polygon has:

   (1) 12 sides (3) 30 sides
   (2) 28 sides (4) cannot be determined

   _____

5. In quadrilateral ABCD m $\angle$ A = 90° m $\angle$ C = 50° and m $\angle$ B = m $\angle$ D. Find the degree measure of $\angle$ B.

   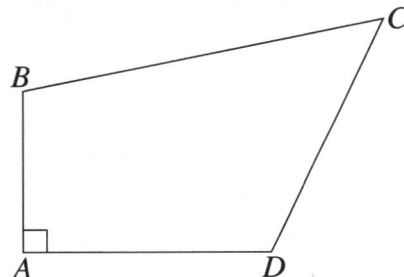

   _____

6. What is the sum of the exterior angles of the polygon below?

   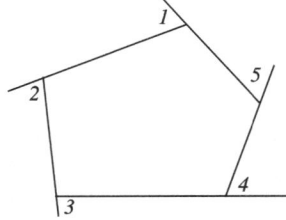

   _____

Chapter 3

MEASUREMENT IN POLYGONS

Interior and Exterior Angles of Regular Polygons

REMEMBER

The measure of each *interior* angle of a regular polygon is equal to $\dfrac{180°(n-2)°}{n}$ where n is the number of sides

The measure of each *exterior* angle of a regular polygon is $\dfrac{360°}{n}$ where n is the number of sides

Example 1: Find the measure of an interior angle of a regular dodecagon (12 sides)
$$\dfrac{180°(n-2)}{n} = \dfrac{180°(12-2)}{12} = \dfrac{180°(10)}{12} = \dfrac{1800}{12} = 150°$$

Example 2: Find the measure of an exterior angle of a regular STOP SIGN
$$\dfrac{360°}{n} = \dfrac{360°}{6} = 60°$$

---

1. How many degrees are there in each interior angle of a regular pentagon?

   (1) 36°   (3) 108°
   (2) 72°   (4) 144°

   _____

2. Each interior angle of a regular polygon measures 140°. How many sides does the polygon have? Explain your answer.

3. In a given regular polygon, the ratio of the number of degrees in an interior angle to the number of degrees in an exterior angle is 7:2. How many sides in this polygon?

   _____

4. Determine whether of not a regular polygon can have an interior angle of 100°. Explain your answer.

5. What is the measure of an exterior angle of a ten sided polygon?

   _____

6. Find the number of degrees in a central angle of the regular hexagon below.

   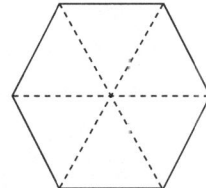

   _____

Chapter 3

MEASUREMENT IN POLYGONS

Parallelograms

> **REMEMBER**
>
> A parallelogram is a quadrilateral with opposite sides parallel. In addition to being parallel, the opposite sides are congruent, the opposite angles are congruent, the consecutive angles are supplementary and the diagonals bisect each other.

1. Using the properties of a parallelogram, fill in the blanks for parallelogram ABCD.

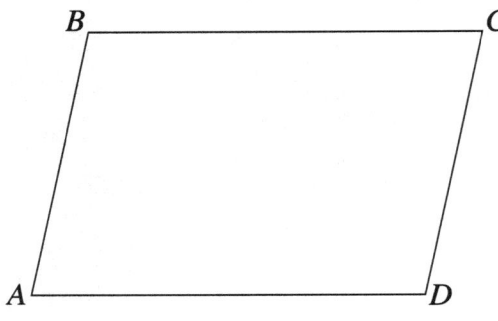

   $\overline{AB}$ ∥ _____

   $\overline{BC}$ ≅ _____

   ∠A ≅ _____

   m∠D + m∠ ___ = 180°

   ∠B ≅ _____

2. In the parallelogram, pictured below $\overline{AE}$ = 3x - 4 and $\overline{EC}$ = x + 12. Find the measure of $\overline{AC}$.

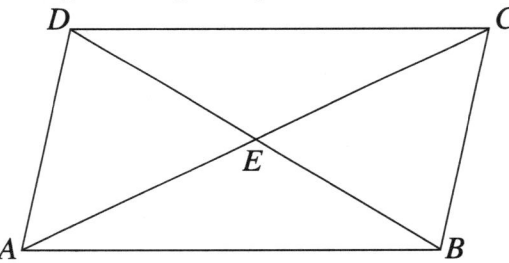

3. In parallelogram DEFG, the measures of ∠D and ∠E are in the ratio 1:4. Find the measure of ∠F in degrees.

4. In parallelogram ABCD, $\overline{AB}$ = (6x - 2) feet, $\overline{DC}$ = (3x + 13) feet and $\overline{AD}$ = 5 feet. Find the number of feet in the perimeter of the parallelogram.

5. In parallelogram DEFG, m∠D = (5x - 10)° and the m∠F = (3x + 6)°. Find the number of degrees in m∠G.

48

Chapter 3

MEASUREMENT IN POLYGONS

Special Parallelograms

REMEMBER

Each of the following are special rectangles with the **same properties as parallelograms** with the **following additional special properties**

Rectangle: a all angles are right angles and congruent and
b the diagonals are congruent

Rhombus: a the sides are congruent
b the diagonals are ⊥ to each other

Square: a all angles are right angles and ≅
b the diagonals are ≅ and ⊥ to each other

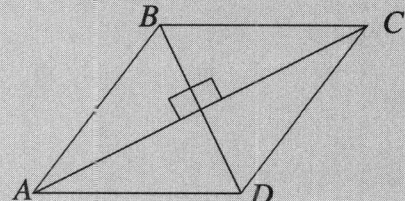

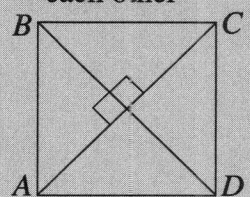

1. In rectangle DEFG, diagonal $\overline{DF}$ = (7x - 10) and diagonal $\overline{EG}$ = 25. Find the value of x.

2. In rectangle ABCD, $\overline{BC}$ = 5 and $\overline{AB}$ = 12. If diagonals $\overline{AC}$ and $\overline{BD}$ bisect at E, find the measure of $\overline{BE}$.

3. The diagonals of rectangle DEFG intersect at H. If $\overline{EH}$ = (2x - 4) and $\overline{HF}$ = (x + 8), then the length of $\overline{DF}$ is:

   (1) 12   (3) 24

   (2) 20   (4) 40

4. In parallelogram SPQR, the ratio of ∠S to ∠P is 1:1. Explain why the parallelogram is a rectangle.

5. In rectangle ABCD, $\overline{AB}$ = 7 and $\overline{BC}$ = 24. Find the measure of diagonal $\overline{BD}$.

6. The diagonals of a rectangle must always be perpendicular to each other. Is this statement true or false? Explain your answer.

49

Chapter 3

MEASUREMENT IN POLYGONS

Midpoints and Centroids in Triangles

REMEMBER

The line jointing the midpoints of two sides of a triangle is equal to ½ of the 3rd side and is parallel to the 3rd side.

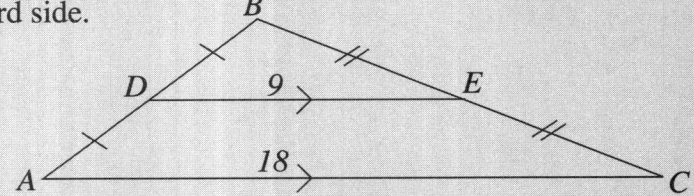

If m$\overline{AC}$ = 18 inches, then m$\overline{DE}$ which connects the midpoint of $\overline{AB}$ to the midpoint of $\overline{BC}$ = 9 inches.

The centroid of a triangle is the point inside the triangle where the medians intersect. The measure of each median segment is in the ratio 2:1

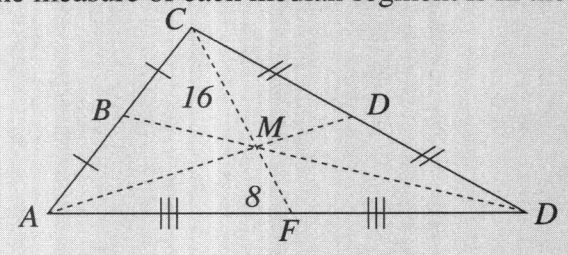

The measure of each median segment is in the ratio 2:1. If $\overline{CF}$ = 24, then $\overline{CM}$ = 16 and $\overline{MF}$ = 8.

---

1. In triangle ABC, $\overline{DE}$ is drawn connecting the midpoint of $\overline{AB}$ to the midpoint of $\overline{BC}$. If m$\overline{AC}$ = 38, find m$\overline{DE}$.

   _____

3. In triangle ACE, $\overline{ED}$ is the median to $\overline{AC}$, $\overline{CF}$ is the median to $\overline{AE}$ and $\overline{AH}$ is the median to $\overline{CE}$. The medians meet at Q. If the m$\overline{AQ}$ is 24, find the m$\overline{AH}$.

   _____

2. In triangle DEF below, C is the centroid. The m$\overline{DC}$ is represented by 2x, m$\overline{CJ}$ is represented by x and m$\overline{DJ}$ = 51. Find m$\overline{DC}$.

   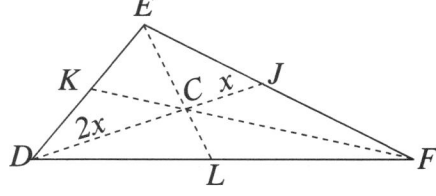

   _____

4. In triangle ABC below, D is the midpoint of $\overline{AB}$ and E is the midpoint of $\overline{BC}$. The m$\overline{AB}$ = 10, m$\overline{BC}$ = 18 and the perimeter triangle ABC = 48. Find the m$\overline{DE}$.

   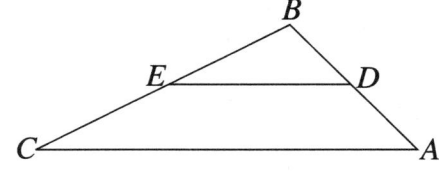

   _____

5. In triangle ABC, D is the midpoint of $\overline{AB}$, E is the midpoint of $\overline{BC}$, F is the midpoint of $\overline{AC}$ and the m$\overline{AF}$ = 14. What is the measure of $\overline{DE}$?

   _____

Chapter 3

MEASUREMENT IN POLYGONS

Right Triangle Proportions

REMEMBER

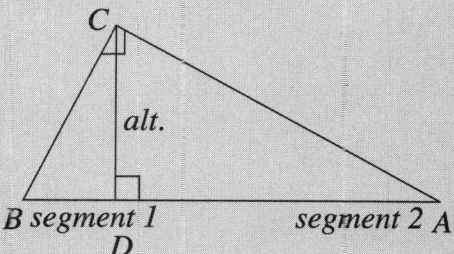

altitude^2 = segment 1 × segment 2

Example 1: In right triangle ABC, the altitude $\overline{CD}$ is drawn to the hypotenuse $\overline{AB}$. If AD = 8 and DB = 2, find CD.

Solution 1: alt^2 = seg 1 × seg 2
alt^2 = 2 × 8 = 16
altitude CD = 4

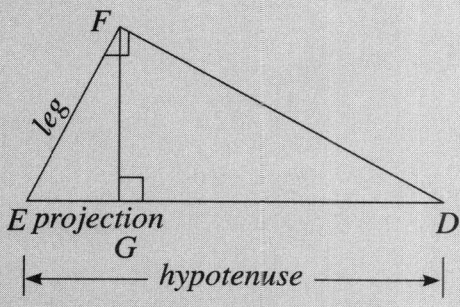

leg^2 = hypotenuse × projection

Example 2: In right triangle DEF, the altitude $\overline{FG}$ is drawn to the hypotenuse $\overline{DE}$. If EG = 4 and hypotenuse DE = 16, find leg EF.

Solution 2: leg^2 = hypotenuse × projection
leg^2 = 16 × 4 = 64
leg EF = 8

---

1. In triangle ABC, angle C is a right angle. Altitude $\overline{CD}$ is drawn to the hypotenuse $\overline{AB}$. BD = 2, DA = 30. Find leg CB.

    (1) $2\sqrt{15}$ (3) 60
    (2) 8   (4) 64

---

2. In triangle DEF, angle F is a right angle. Altitude $\overline{FG}$ is drawn to the hypotenuse $\overline{DE}$. EG = 20, GD = 5. Find altitude FG.

    (1) $5\sqrt{5}$ (3) $10\sqrt{5}$
    (2) 10   (4) 100

---

3. In triangle ABC, angle C is a right angle. Altitude $\overline{CD}$ is drawn to the hypotenuse $\overline{AB}$. CD = 12. The lengths of the segments of the hypotenuse *cannot* be:

    (1) 3 and 48   (3) 10 and 14.4
    (2) 7 and 18   (4) 12 each

Chapter 3

MEASUREMENT IN POLYGONS

Right Triangle Proportions (continued)

4. In triangle DEF, angle F is a right angle. Altitude $\overline{FG}$ is drawns to the hypotenuse $\overline{DE}$. DG = 14, the hypotenuse DE = 18. Find the value of leg EF.

   (1) $6\sqrt{2}$ (3) $6\sqrt{7}$

   (2) $14\sqrt{2}$ (4) $2\sqrt{14}$

   _____

5. In right triangle ABC, angle C is a right angle. Altitude $\overline{CD}$ bisects hypotenuse $\overline{AB}$ at D. AB = 50. Show how you arrived at your answers to:

   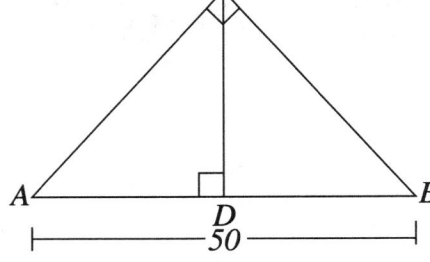

   a   Find altitude CD
   b   Find leg AC in simplest radical form.
   c   Find leg AC to the nearest integer.

   a _____   *b* _____   *c* _____

6. Find the measure of the segment marked x in the triangle pictured.

   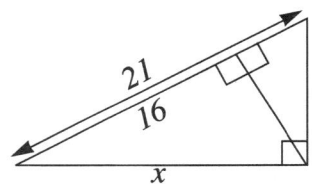

   _____

7. In triangle ABC, $\overline{AB}$ is the hypotenuse and $\overline{CD}$ is the altitude to $\overline{AB}$. If CD = 6 and $\overline{DB}$ exceeds $\overline{AD}$ by 5, find $\overline{AD}$ and $\overline{DB}$.

   _____

8. The measure of x in the triangle pictured is:

   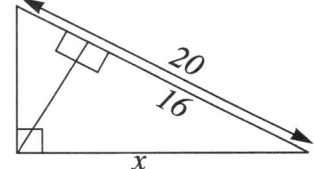   (1) 8.0   (3) 17.8

                                    (2) 8.1   (4) 17.9

   _____

Chapter 3

MEASUREMENT IN POLYGONS

Proportional Sides of a Triangle

REMEMBER

If in triangle ABC, $\overline{DE} \parallel \overline{AC}$, then $\overline{AD} : \overline{DB} = \overline{EC} : \overline{BE}$ and $\overline{DE} : \overline{AC} = \overline{DB} : \overline{AB}$.
The perimeters of the triangles are also in proportion.

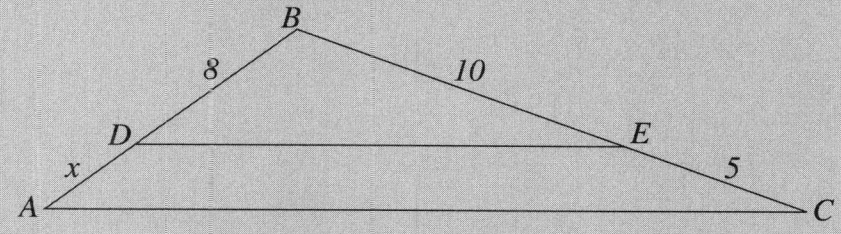

Example: If mAD = x, the mDB = 8, the mEC = 5, mBE = 10, find mAD.

Solution: $\dfrac{x}{8} = \dfrac{5}{10}$

10x = 40, so x = AD = 4

1. In triangle ABC, DE ∥ AC. If BD = 40, DA = 60 and AC = 150, find mDE.

2. In triangle ABC, DE ∥ to AB. If DE is one-fifth AB, what is the ratio of the area of triangle DEC to the area of triangle ABC?

3. In the accompanying figure, segment BD is perpendicular to segment DC in triangle BDC, and segment EF is perpendicular to segment BD at point E. If DC = 7, BD = 22, and BE = 11.

    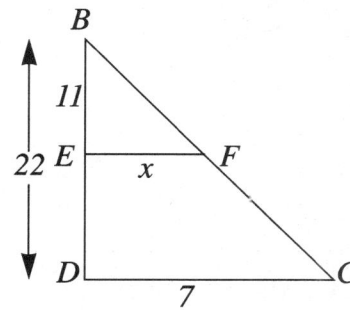

    a   find segment EF

    b   Why is EF parallel to DC?

    c   Why is point F a midpoint on segment BC?

    a _____   b _____   c _____

4. In the diagram, the nonparallel sides of trapezoid ABCD are extended to point X. AB = 5, DC = 15 and AD = 10.
    a   Find AX
    b   If trapezoid ABCD were isosceles, is triangle XAB isosceles? Explain.
    c   Find the median of trapezoid ABCD.

    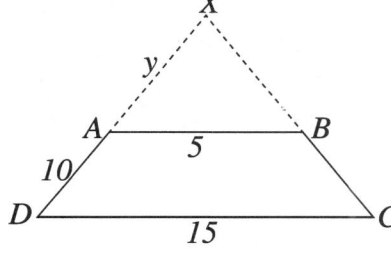

    a _____   b _____   c _____

53

Chapter 3

MEASUREMENT IN POLYGONS

Pythagorean Theorem

REMEMBER

RIGHT TRIANGLE: $Leg^2 + Leg^2 = Hypotenuse^2$
$a^2 + b^2 = c^2$

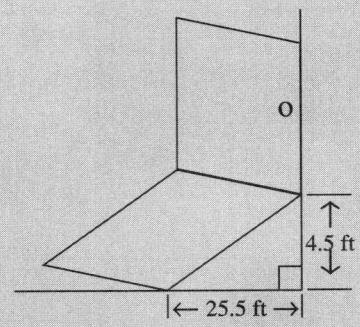

Example: A diner requires a concrete ramp to make its entrance accessible to the physically challenged. The entrance is 4.5 feet above the walkway. A ramp must be built with a horizontal span of 25.5 feet. Find the length of the ramp to the nearest foot.

Solution: $a^2 + b^2 = c^2$ $\quad 4.5^2 + 25.5^2 = c^2$ $\quad 20.25 + 650.25 = c^2$ $\quad c = 26$ ft.

1. If the legs of a right triangle are 5 and 9, what is the length of the hypotenuse?

    (1) 14 $\qquad$ (3) $\sqrt{56}$

    (2) $\sqrt{14}$ $\qquad$ (4) $\sqrt{106}$

2. The length of a side of a square is 7. In simplest radical form, find the length of a diagonal of the square.

    (1) $\sqrt{14}$ $\qquad$ (3) $2\sqrt{7}$

    (2) $7\sqrt{2}$ $\qquad$ (4) $49\sqrt{2}$

3. In rectangle ABCD, AB = 15 and diagonal BD = 39. What is the *perimeter* of ABCD?

    (1) 51 $\qquad$ (3) 540

    (2) 102 $\qquad$ (4) 585

4. The length of the hypotenuse of a right triangle is 4 centimeters more than the longer leg. The length of the longer leg is 14 centimeters more than the length of the shorter leg. Find the number of centimeters in the length of each side of the right triangle. Show how you arrived at your answer.

Chapter 3

MEASUREMENT IN POLYGONS

Pythagorean Theorem (continued)

5. A builder is putting up two walls of a house that must be 90° to each other. The first wall is 15 feet long. He levels the wall and nails it in place. The second wall is 8 feet long. He figures that the two walls can be considered as legs of a right triangle and that the distance between the two unattached wall ends can be the hypotenuse. What distance must this be if the two walls are to be meeting at 90°? Show how you arrived at your answer.

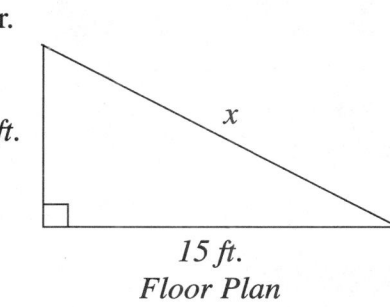

Floor Plan

6. Mike is building a 10 ft by 16 ft shed. He wants to make sure the sides make an angle of 90° with each other. He marks off 5 feet on the 10 ft walls and 12 feet on the 16 ft walls. What value does Mike expect to get for x? Show how you arrived at your answer.

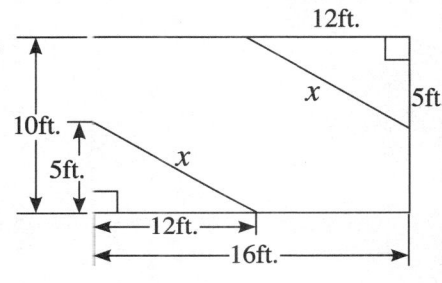

Floor Plan

7. The hypotenuse of a right triangle is represented by 2x - 3 and one leg is represented by x. The other leg is 12. Find: *a*. hypotenuse, *b*. perimeter, and *c*. area of the right triangle. Show how you arrived at your answers.

a _____   b _____   c _____

Chapter 3

MEASUREMENT IN POLYGONS

Trapezoids

> **REMEMBER**
>
> A trapezoid is a quadrilateral that has two and only two sides parallel. The parallel sides are called the bases and the non-parallel sides are called the legs. If the two legs are equal in length then the trapezoid is an isosceles trapezoid.
>
> The median in a trapezoid is parallel to the two bases and equal to one-half their sum.
>
> **Example 1:** If $\overline{RT}$ is a median of trapezoid ABCD. Find the length of $\overline{RT}$.
>
>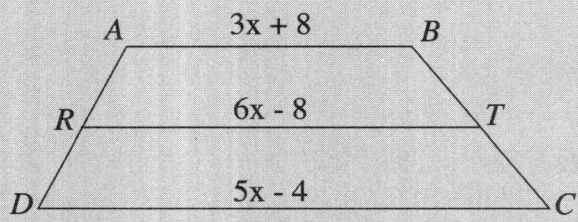
>
> **Example 2:** In isosceles trapezoid ABCD. $\overline{AB} \parallel \overline{DC}$. If $\overline{AD} = 3x + 4$ and $\overline{BC} = x + 12$, find x.
>
>
>
> **Solution 1:** $RT = \frac{1}{2}(\overline{AB} + \overline{CD})$
>
> $6x - 8 = \frac{1}{2}(3x + 8 + 5x - 4)$
>
> $6x - 8 = \frac{1}{2}(8x + 4)$
>
> $6x - 8 = 4x + 2$
> $2x = 10$
> $x = 5$ so $\overline{RT} = 6x - 8 = 22$
>
> **Solution 2:** $\overline{AD} \cong \overline{BC}$
> $3x + 4 = x + 12$
> $2x = 8$
> $x = 4$
>
> **Example 3:** The area of a trapezoid with parallel bases b and B and height h is given by the formula $A = \frac{1}{2}h(b + B)$. If $h = 10$, $b = 20$ and $B = 30$, then $A = 250$

1. In the accompanying diagram, isosceles trapezoid CDEF has bases of lengths 6 and 12 and an altitude of length 4. Find $\overline{CD}$.

   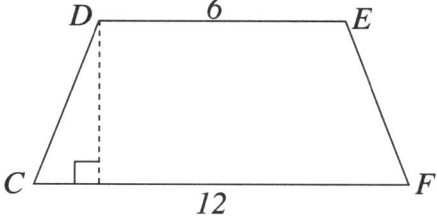

2. In the diagram of trapezoid ABCD, CD = 10 m∠A = 45°, m∠D = 90° and base BC = 3. Find the length of base $\overline{AD}$.

   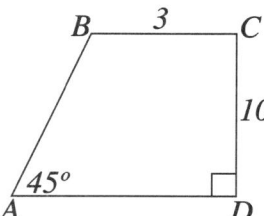

3. In trapezoid ABCD, $\overline{AB} \parallel \overline{CD}$. If AB = 2 and CD = 50, what is the measure of the median $\overline{MN}$?

Chapter 3

MEASUREMENT IN POLYGONS

Trapezoids (continued)

4. Find the area of trapezoid ABCD given that the area of right triangle ABE is 30 in.², DC = 14", AE = 6".

   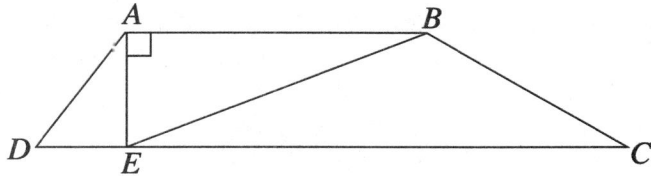

5. Find the length of the median of a trapezoid if the lengths of the two bases are 20 and 46 inches.

6. In isosceles trapezoid RSTU, RS = 8x + 5 and TU = 2x + 29. Express the length of the median $\overline{QP}$ in terms of x.

   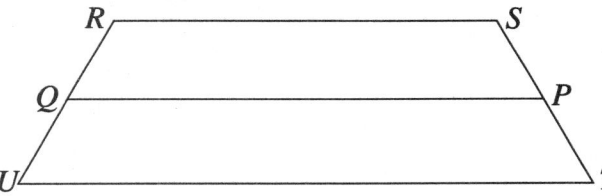

7. In isosceles trapezoid ABCD, $\overline{AB} \parallel \overline{CD}$, AB = 18, CD = 26 and AD = 5. Find the length of an altitude of ABCD.

   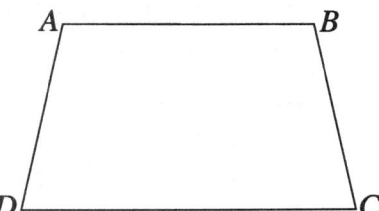

8. In isosceles trapezoid DEFG, DE = 20, FG = 30 and DH = 12. Find the length of $\overline{GD}$.

   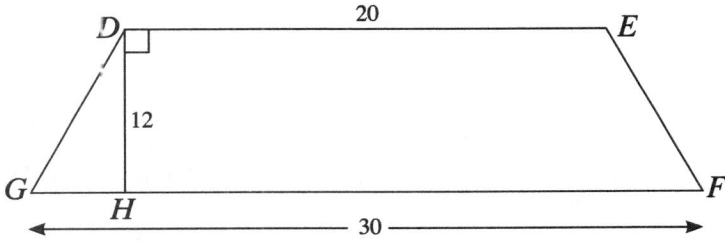

Chapter 4

MEASUREMENT IN CIRCLES

Chords, Arcs and Perpendicular Bisectors

> **REMEMBER**
>
> In the same or equal circles, equal chords are equally distant from the center and if the chords are equally distant from the center they are equal in length.
>
> In the same **or** equal circles, equal arcs have equal chords and equal chords have equal arcs.
>
> A line passing through the center of a circle and perpendicular to a chord, bisects the chord and its arcs.

1. In $\odot O$, $\overline{AB} \perp \overline{CD}$, $OC = 17$ and $CE = 15$. Find $\overline{OE}$.

   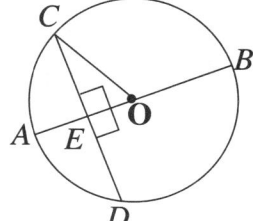

2. In $\odot O$, m $\overline{BC}$ = m $\overline{AD}$ and M is the midpoint of $\overline{BC}$. If $BM = 8$, find the m $\overline{AD}$.

   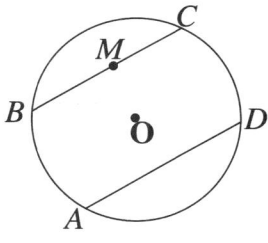

3. In $\odot O$, diameter $\overline{AB} \perp \overline{CD}$. If $DC = 20$, find $\overline{DE}$.

   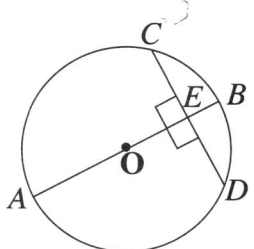

4. In $\odot O$, if $\overline{OA} \perp \overline{CD}$, $\overline{OB} \perp \overline{EF}$, $\overline{OA} \cong \overline{OB}$ and $CD = 13$. Find $\overline{EF}$.

   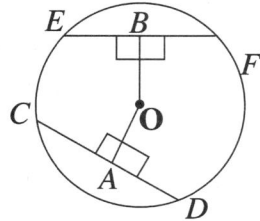

5. In $\odot O$, $\overline{ON} \perp \overline{AB}$, $AB = 24$, $OM = 5$. Find x.

   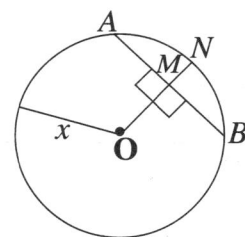

6. In $\odot O$, m$\widehat{AMC} = 240°$. If $\overline{AB} \cong \overline{BC}$, find m $\widehat{AB}$

   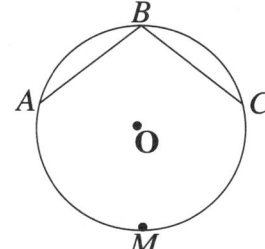

Chapter 4

MEASUREMENT IN CIRCLES

Chords, Arcs and Perpendicular Bisectors (continued)

7. $\overline{AB}$ is a diameter in $\odot O$. If $\overline{AB} \perp \overline{CD}$ and $\overline{AB} \perp \overline{EF}$,
   a. Explain why $\overline{EF} < \overline{CD}$.
   b. What happens to the central angle as the chord increases in length?

   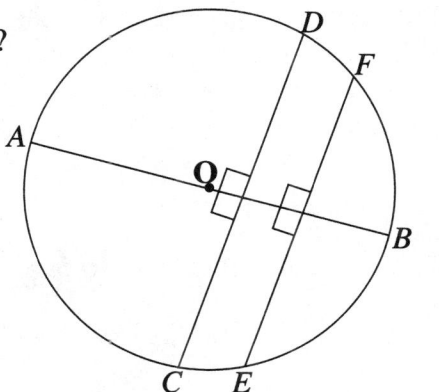

   a _____

   b _____

8. In $\odot O$, $\overline{AB} \cong \overline{CD}$ and $\overline{MN}$, $\overline{AD}$ and $\overline{BC}$ are diameters. If m∠CON = 25° and m∠DON = 55°, find m∠AOB.

   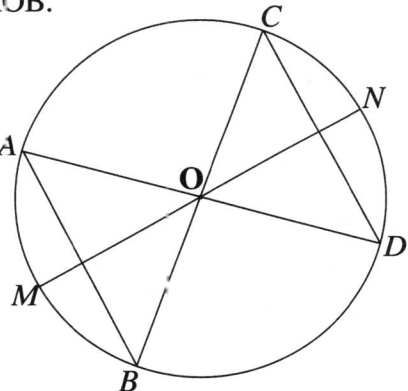

   _____

9. In $\odot O$, $\overline{AB} \parallel \overline{CD}$. If $\widehat{AB} = 75°$ and $\widehat{CD} = 95°$, find m$\widehat{AC}$.

   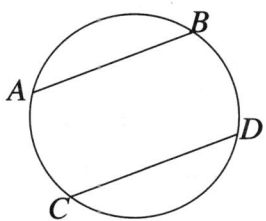

   _____

10. In $\odot O$, $\overline{AB} \parallel \overline{CD}$. If m$\widehat{AC} = 58°$ and $\overline{AC} = 9$, find $\overline{BD}$.

    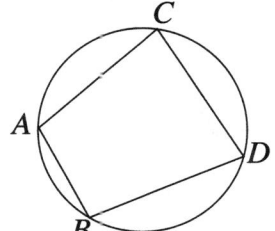

    _____

Chapter 4

MEASUREMENT IN CIRCLES

Central and Inscribed Angles

> **REMEMBER**
>
> **The central angle of a circle has the same measure as its intercepted arc.**
>
> **Example:** Find the measure of central ∠ AOB if major $\widehat{AB}$ = 290°
> **Solution:** Degrees in ⊙O - degrees in major $\widehat{AB}$ =
> degrees in minor $\widehat{AB}$ = 360° - 290° = 70° so
> m ∠ AOB = minor $\widehat{AB}$ = 70°
>
>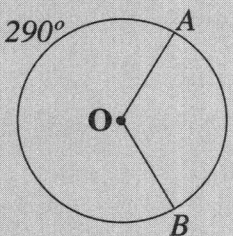
>
> **An inscribed angle in a circle = $\frac{1}{2}$ of its intercepted arc.**
>
> **Example:** Find the measure of inscribed ∠ ABC if $\widehat{AB}$ = 140° and $\widehat{BC}$ = 120°.
> **Solution:** Degrees in ⊙O - degrees in ($\widehat{AB}$ + $\widehat{BC}$) =
> 360° - (140° + 120°) = $\frac{1}{2}$(100)° = 50°
>
>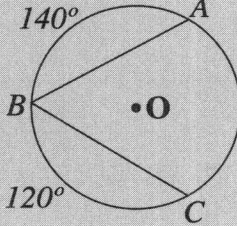

1. In the accompanying diagram, $\overline{BC}$ is a diameter and $\widehat{AC}$ = 120°. How many degrees are there in m ∠ ACB?

   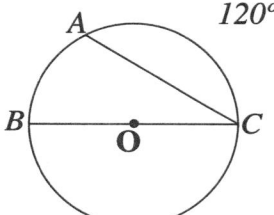

   _____

2. In the accompanying figure of circle O, radii $\overline{OC}$ and $\overline{OD}$ are drawn. If m ∠ D = 44°, the m$\widehat{CD}$ is

   (1) 92° (3) 88°
   (2) 46° (4) 44°

   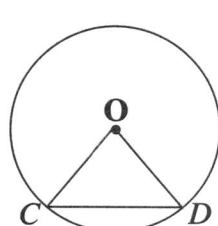

   _____

3. In the diagram below, find the measure of ∠ ABC if the m$\widehat{AC}$ = 92°

   (1) 92° (3) 23°
   (2) 184° (4) 46°

   ![diagram]

   _____

4. In the accompanying diagram, isosceles △ ABC is inscribed in the circle. If $\overline{AB} \cong \overline{CB}$ and m$\widehat{AB}$ = 145°, find m ∠ B.

   ![diagram]

   _____

5. In the diagram of circle O, radii $\overline{OA}$, $\overline{OB}$ and $\overline{OC}$ are drawn. If m ∠ AOB = 160° and m$\widehat{BC}$ = 70°, find m ∠ AOC.

   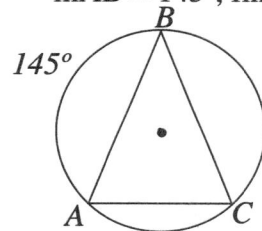

   _____

6. In the accompanying diagram of circle O, m ∠ ABC = 60°. Find m ∠ AOC.

   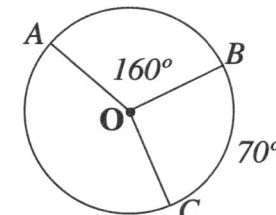

   _____

Chapter 4

MEASUREMENT IN CIRCLES

Angles Formed by a Tangent and a Chord

REMEMBER

An angle formed by a tangent to a circle and a chord intersecting at the point of tangency is equal to one-half the measure of the intercepted arc.

Example: Find the measure of $\angle$ x in the diagram below if $\overline{AB}$ is a chord, $\overleftrightarrow{BT}$ is a tangent to circle O at B, and major $\overparen{AB} = 290°$.

Solution: Degrees in a circle: 360°
Measure of major $\overparen{AB} = 290°$
Since the measure of the intercepted arc $\overparen{AB}$ is 70°,
$m \angle x = 35°$ ans.

1. In circle O, $\overline{BC}$ is a chord and $\overleftrightarrow{CD}$ is a tangent. Find the value of x.

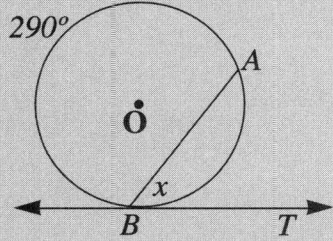

2. In circle O, $\overline{AB}$ is a chord and $\overleftrightarrow{BR}$ is a tangent. Find the value of x and y.

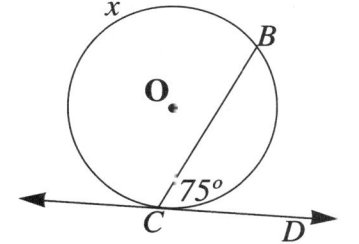

x = _____

y = _____

3. A regular pentagon RSTUV is inscribed in a circle. Find the measure of the acute angle formed by side $\overline{ST}$ and the tangent at T.

(1) 72°   (3) 144°
(2) 36°   (4) 18°

_____

4. In the accompanying diagram, $\overleftrightarrow{BD}$ is tangent to circle O at B, $\overline{BC}$ is a chord, and $\overline{BOA}$ is a diameter. If m$\overparen{AC}$:m$\overparen{CB}$ = 1:5, find the m $\angle$ DBC.

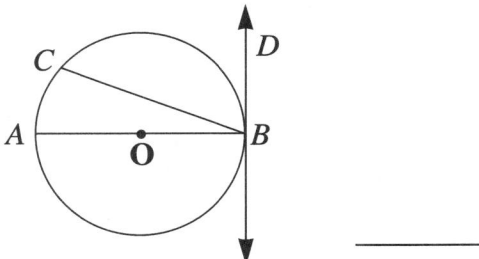

_____

5. In circle O, $\overline{AB} \cong \overline{AC} \cong \overline{BC}$ and $\overleftrightarrow{CD}$ is a tangent at C. Find m $\angle$ BCD.

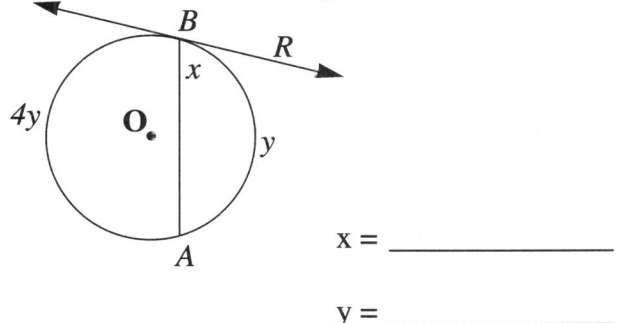

_____

6. Triangle DEF is inscribed in circle O. m$\overparen{DE}$:m$\overparen{EF}$:m$\overparen{DF}$ = 2:3:4. Find the measure of the acute angle formed by side $\overline{EF}$ and the tangent to the circle at F.

(1) 40°   (3) 80°
(2) 60°   (4) 100°

_____

61

Chapter 4

MEASUREMENT IN CIRCLES

Angles Formed by Two Chords

> **REMEMBER**
>
> An angle formed by two chords intersecting within a circle is equal to one-half the measure of the sum of the arcs intercepted by the angle and its vertical angle.
>
> **Example:** In circle O, $\overline{DC}$ is a chord and $\overline{AB}$ is a diameter. If $m\widehat{AC} = 80°$ and $m\widehat{AD} = 50°$, find the value of x.
>
> **Solution:**
> $m\widehat{ACB} = 180°$ (diameter bisects the circle)
> $m\widehat{AC} = 80°$
> $m\widehat{CB} = 100°$
> $m\angle x = \frac{1}{2}(m\widehat{CB} + m\widehat{AD})$
> $m\angle x = \frac{1}{2}(100 + 50)$
> $m\angle x = \frac{1}{2}(150)$
> $m\angle x = 75°$ ans.

1. In the accompanying diagram, chords $\overline{AB}$ and $\overline{CD}$ intersect at E. If $m\angle AED = 104°$ and $m\widehat{CB} = 86°$, find the $m\widehat{AD}$.

   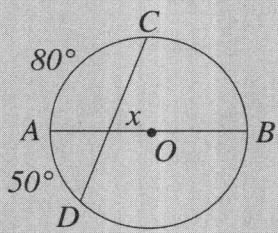

   (1) 122°   (3) 95°
   (2) 52°   (4) 86°

2. In circle O, diameter $\overline{AB} \perp \overline{CD}$. If $m\widehat{AD} = 110°$, find the $m\widehat{BC}$.

   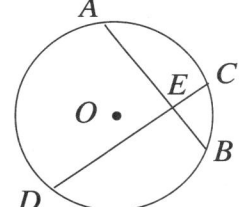

3. In circle O, chords $\overline{AB}$ and $\overline{CD}$ intersect at E. If $m\widehat{BC} = 50°$ and $m\widehat{AD} = 90°$, find the $m\angle AEC$.

   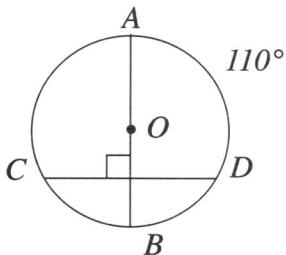

   (1) 70°   (3) 135°
   (2) 45°   (4) 110°

4. In the diagram below, chords $\overline{AB}$ and $\overline{CD}$ intersect at E. If $m\angle AEC = 3x°$, $m\widehat{AC} = 160°$, and $m\widehat{DB} = x°$, what is the value of x?

   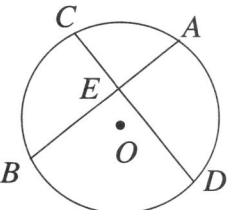

   (1) 40°   (3) 32°
   (2) 96°   (4) 80°

5. Two chords intersecting within a circle form an angle whose measure is 70°. If one of the intercepted arcs meaures 90°, what is the measure of the other intercepted arc?

6. In a circle, chords $\overline{AB}$ and $\overline{CD}$ are perpendicular, and intersect at E. If $m\widehat{AC} = 70°$, fing the $m\widehat{BD}$.

62

Chapter 4

MEASUREMENT IN CIRCLES

Angles Formed by Two Tangents, Two Secants or a Tangent and a Secant

> **REMEMBER**
>
> Angles formed by a tangent and a secant, or two secants, or two tangents intersecting outside a circle have one-half the measure of the difference of the intercepted arcs.
>
> **Example:** In circle O, m$\widehat{AE}$ = 100° and m∠C = 20°, Find the m$\widehat{BD}$.
>
>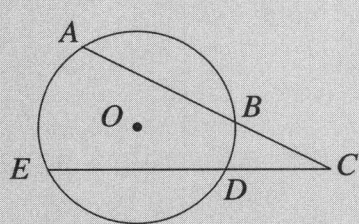
>
> **Solution:**
> Let m$\widehat{BD}$ = x
>
> m∠C = $\frac{1}{2}$(m$\widehat{AE}$ − m$\widehat{BD}$)
>
> 20 = $\frac{1}{2}$(100 − x)
>
> 40 = 100 − x
>
> x = 60°   ans.

1. Two tangents to a circle from an external point intercept a major arc of 280°. Find the number of degrees in the angle formed by the two tangents.

   _____

2. In the accompanying diagram, tangent $\overline{PA}$ and secant $\overline{PBC}$ are drawn to circle O. If m$\widehat{AC}$ is four times m$\widehat{AB}$ and m∠P = 60°, find m$\widehat{AB}$.

   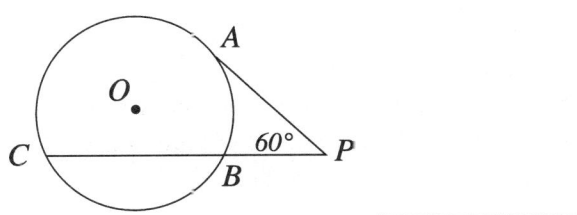

   _____

3. In the diagram below, $\overline{PA}$ and $\overline{PB}$ are tangents to circle O. Find the value of x.

   (1) 135°   (3) 50°
   (2) 85°   (4) 15°

   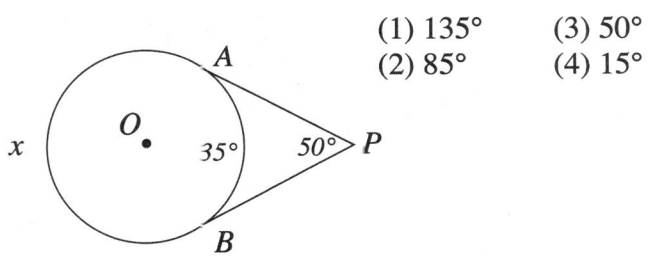

4. In the accompanying diagram, $\overline{PBA}$ and $\overline{PCD}$ are secants to the circle. If m$\widehat{AD}$ = 140°, and m$\widehat{BC}$ = 60°, find m∠P.

   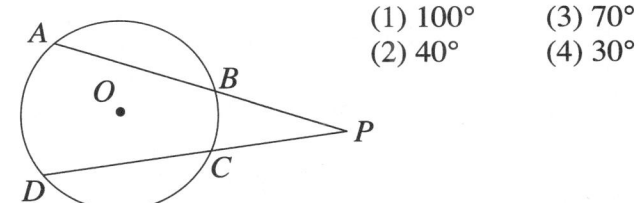

   (1) 100°   (3) 70°
   (2) 40°   (4) 30°

   _____

5. In the accompanying diagram, tangent $\overline{PA}$ and secant $\overline{PBC}$ are drawn to circle O from point P. If m$\widehat{AC}$ = 110° and m∠P = 20°, find the m$\widehat{AB}$.

   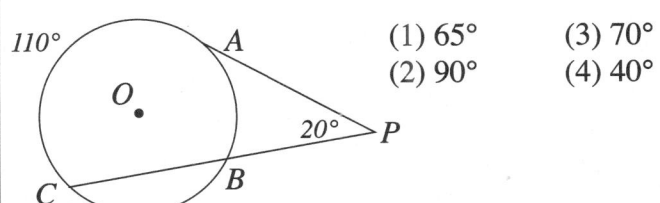

   (1) 65°   (3) 70°
   (2) 90°   (4) 40°

   _____

6. In the diagram below, secants $\overline{CBA}$ and $\overline{CDE}$ are drawn to circle O from point C. Find the value of x.

   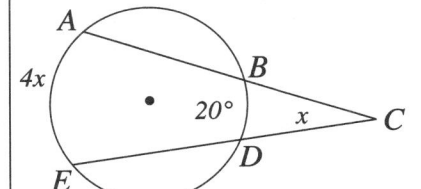

   _____

63

Chapter 4

MEASUREMENT IN CIRCLES

Angles in Circles, Combination

1. If $\overline{PC}$ is a secant and $\overline{PA}$ is a tangent in $\odot$ O, m $\widehat{AB}$ = 70°, find m $\angle$ PAC.

   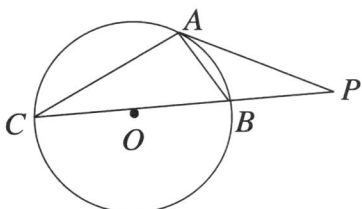

   _____

2. If $\triangle$ ABC is inscribed in $\odot$ O, m $\angle$ AEB = 55° and m$\widehat{EC}$ = 120°, find m $\angle$ ADB.

   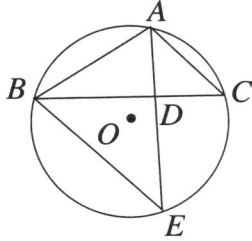

   _____

3. In the accompanying diagram of circle O, $\overline{PBA}$ and $\overline{PCD}$ are secants, chords $\overline{AC}$ and $\overline{BD}$ intersect at E, $\overline{BA} \cong \overline{CD}$, chord $\overline{BC}$ is drawn, m $\angle$ ABD = 65°, and m$\widehat{BC}$ = 60°.

   Find: a m $\angle$ ACD   d m $\angle$ AED
          b m $\angle$ P   e m $\angle$ PCB
          c m $\angle$ DBC

   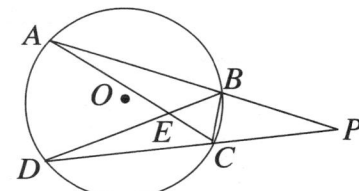

   a_____ b_____ c_____ d_____ e_____

4. In the accompanying diagram of circle O, $\overline{AE}$ and $\overline{FD}$ are chords $\overline{AOBG}$ is a diameter and is extended to C. $\overline{CDE}$ is a secant, $\overline{AE} \parallel \overline{FD}$, and m$\widehat{AE}$: m$\widehat{ED}$: m$\widehat{DG}$ = 3 : 2 : 1.

   Find: a m$\widehat{DG}$ d m $\angle$ DCA
          b m $\angle$ AEF   e m $\angle$ CDF
          c m $\angle$ DBG

   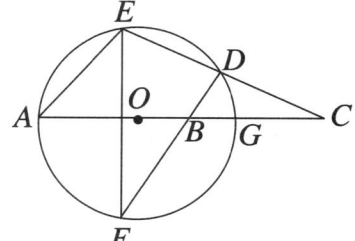

   a_____ b_____ c_____ d_____ e_____

64

Chapter 4

MEASUREMENT IN CIRCLES

Angles in Circles, Combination (continued)

5. In the accompanying diagram of circle O, $\overrightarrow{EA}$ is a tangent, $\overline{EBC}$ is a secant, D is the midpoint of $\overparen{AC}$, m$\overparen{AD}$ = 86° and m$\overparen{AB}$ = 62°.

    Find: a m$\overparen{BC}$ d m∠AFB
            b m∠E   e m∠EBA
            c m∠ABD

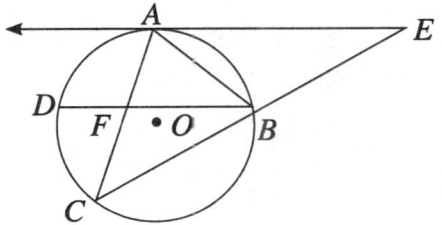

a_____ b_____ c_____ d_____ e_____

6. In ⊙ O, $\overline{ABC}$ and $\overline{ADE}$ are secants, $\overparen{BC} \cong \overparen{ED}$. If m$\overparen{EC}$ = 178° and m$\overparen{BC}$ = m$\overparen{ED}$ = 64°, find m∠A.

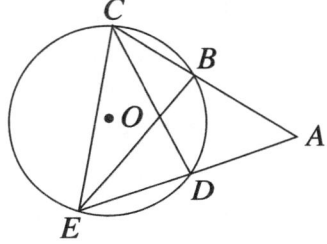

_____

7. In the accompanying diagram, tangent $\overline{PB}$ and secant $\overline{PCA}$ are drawn to ⊙ O. The m$\overparen{AB}$ = 150° and m$\overparen{AC}$ = 120°.
   Find a m$\overparen{BC}$ b m∠PAB   c m∠APB

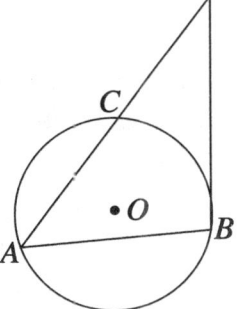

a_____ b_____ c_____

8. In ⊙ O, $\overline{AC}$ is a diameter. $\overrightarrow{CD}$ is tangent to ⊙ O at C. If m$\overparen{AB}$ = 76°, find m∠BCD.

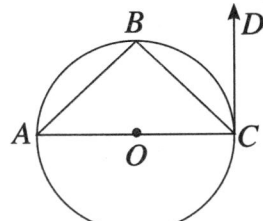

_____

Chapter 4

MEASUREMENT IN CIRCLES

Lengths of Tangents and Secants

> **REMEMBER**
>
> When a tangent and a secant are drawn to a circle from an outside point, the square of the length of the tangent segment is equal to the product of the lengths of the secant and its *external* segment.
>
> **Example:** In the accompanying diagram, $\overrightarrow{PT}$ is tangent to circle O at T and $\overline{PKM}$ is a secant. If PK = 4, and MK = 12, find PT.
>
> **Solution:**
> $(PT)^2 = PM \cdot PK$
> Let PT = x
> $x^2 = 16 \cdot 4$
> $x^2 = 64$
> $x = 8$
> PT = 8   ans
>
> Note: $\overline{PM} \cong \overline{PK} + \overline{KM}$
>
>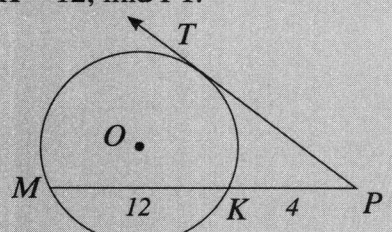

1. In the accompanying figure, $\overrightarrow{PA}$ is tangent to circle O at A and $\overline{PBC}$ is a secant. If PC = 25, and PB = 4, find the length of PA.

   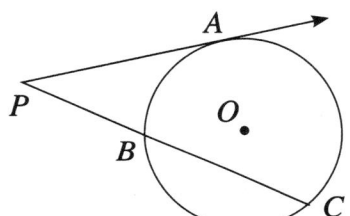

2. In the accompanying diagram, $\overrightarrow{PC}$ is tangent to circle O at C and $\overline{PAB}$ is a secant. If PC = 10, and AB = 21, find PA.

   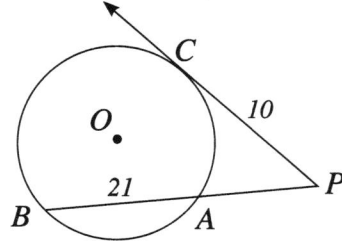

3. In the accompanying diagram, $\overrightarrow{PA}$ is tangent to circle O at A and $\overline{PBC}$ is a secant. If PA = 9, and PB = 3, find BC.

   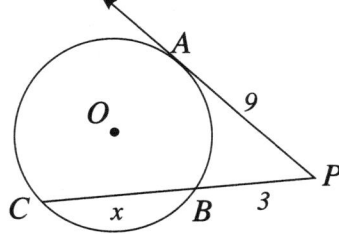

   (1) 27   (3) 3
   (2) 18   (4) 24

4. In the accompanying figure, $\overrightarrow{PA}$ is tangent to circle O at A and $\overline{PBC}$ is a secant. If PB = 4, and BC = 8 find PA.

   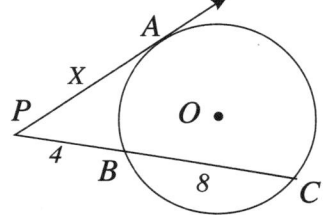

   (1) $2\sqrt{3}$ (3) $4\sqrt{2}$
   (2) $4\sqrt{3}$ (4) $2\sqrt{2}$

5. In the accompanying diagram, $\overrightarrow{PD}$ is tangent to circle O at D and $\overline{PRS}$ is a secant. If PD = x, PR = 12, and RS = x - 3, find the value of x.

   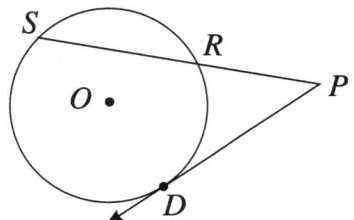

6. In the accompanying figure, $\overrightarrow{AB}$ is tangent to circle O at B and $\overline{ACD}$ is a secant. If DC = 12, and AB = 8, find AC.

   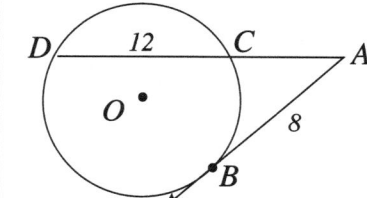

Chapter 4

MEASUREMENT IN CIRCLES

Lengths of Two Intersecting Chords

> **REMEMBER**
>
> When two chords intersect within a circle, the product of the lengths of the segments of one chord is equal to the product of the lengths of the segments of the other chord.
>
> **Example:** In the accompanying figure, chords $\overline{RS}$ and $\overline{TV}$ intersect at P.
> If TV = 10, PV = 4, and RP = 8, find PS.
>
>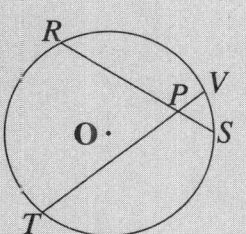
>
> **Solution:**   RP•PS = TP•PV
> Let PS = y   8y = 6•4
>   8y = 24
>   y = 3
>   PS = 3   ans.   Note: $\overline{TP} \cong \overline{TV} - \overline{PV}$

1. In the accompanying figure, chords $\overline{AB}$ and $\overline{CD}$ intersect at E. If CD = 13, EB = 3 and CE = 4, find AE.

   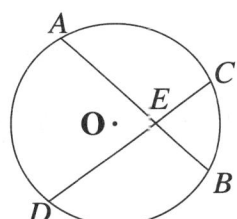

2. In circle O, diameter $\overline{AB}$ is perpendicular to chord $\overline{CD}$ at E. If AE = 24 and EB = 6, what is CD?

   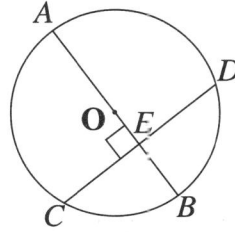

   (1) 12   (3) 18
   (2) 15   (4) 24

3. In circle O, chords $\overline{AB}$ and $\overline{CD}$ intersect at P. If AP = x, PB = y and CP = z, what is the length of PD in terms of x, y and z?

   (1) $\dfrac{xy}{z}$ (3) $\dfrac{yz}{x}$

   (2) $\dfrac{xz}{y}$ (4) $\dfrac{x+y}{z}$

4. In the accompanying diagram, chords $\overline{AB}$ and $\overline{CD}$ of circle O intersect at E. If AE = x, EB = x-8, CE = 5 and ED = 4, find AE.

   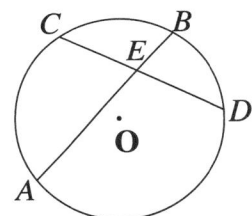

5. In the accompanying diagram, $\overline{AB}$ and $\overline{CD}$ are chords of the circle and intersect at E. If AE = 3, EB = 5, CE = x and ED = x+2, find the value of x.

   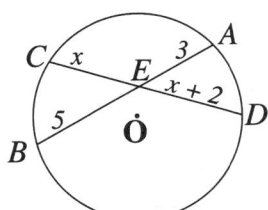

6. In the accompanying figure, diameter $\overline{AB}$ is perpendicular to chord $\overline{CD}$ at E. If CE = 8, and EB = 4, find AE.

   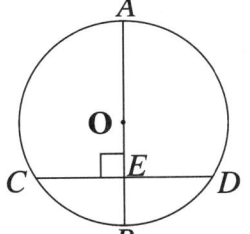

   (1) 10   (3) 16
   (2) 2   (4) 4

67

Chapter 4

MEASUREMENT IN CIRCLES

Lengths of Two Secants

REMEMBER

When two secants are drawn to a circle from an outside point, the product of lengths of one secant and its external segment equals the product of the other secant and its external segment.

Example: In the accompanying diagram, secants $\overline{ABC}$ and $\overline{ADE}$ are drawn to a circle from external point A. If AB = 6, BC = 8 and AE = 21, find AD.

Solution: AE•AD = AC•AB
Let AD = x   21x = 14•6
              21x = 84
              x = 4
              AD = 4   ans.

Note: $\overline{AC} \cong \overline{AB} + \overline{BC}$

1. In the accompanying figure, secants $\overline{PAB}$ and $\overline{PCD}$ are drawn to circle O from P. If PA = 3, AB = 9 and PD = 18, find PC.

   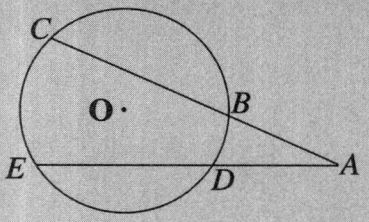

2. Two secants, $\overline{ABC}$ and $\overline{ADE}$ are drawn to a circle from external point A. If AB = 5, BC = 7 and AD = 4, find DE.

   (1) 15   (3) 3
   (2) 14   (4) 11

3. From point P outside of circle O, secants $\overline{PAB}$ and $\overline{PCD}$ are drawn. If PA = x, AB = x+2, PC = 6 and CD = 4, find the value of x.

   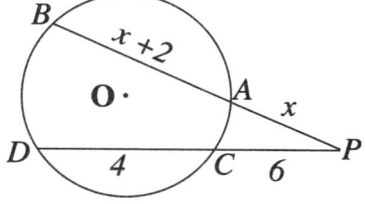

4. In the accompanying diagram, secants $\overline{RST}$ and $\overline{RUV}$ are drawn to circle O from R. If RT = 18, RS = 3, and RV = 27, find VU.

   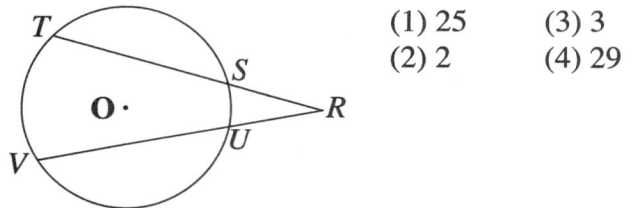

   (1) 25   (3) 3
   (2) 2   (4) 29

5. Two secants, $\overline{ABC}$ and $\overline{ADE}$ are drawn to a circle from external point A. If AB = 7, BC = 1 and AD = 4, find DE

   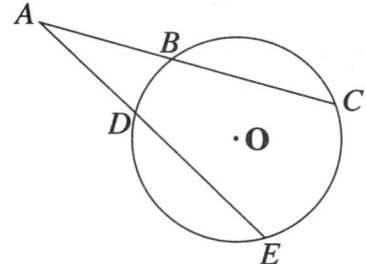

6. Two secants, $\overline{PDQ}$ and $\overline{PBA}$ are drawn to a circle from external point P. If PQ = x, PD = y and PA = z, find PB in terms of x, y and z.

Chapter 4

MEASUREMENT IN CIRCLES

Parallel and Perpendicular Lines in Circles

REMEMBER

A radius or diameter drawn to a tangent in a circle is perpendicular at the point of tangency.

Example: In $\odot O$, $\overline{OA} \perp \overline{PA}$. If $\overline{OA} = 8$ and $OP = 17$, find PA.

Solution: In right triangle OAP, $\overline{OA}$ is a leg and $\overline{OP}$ is the hypotenuse so using $\overline{OA}^2 + \overline{PA}^2 = \overline{OP}^2$, $8^2 + \overline{PA}^2 = 17^2$, $64 + \overline{PA}^2 = 289$, $\overline{PA}^2 = 225$ so $PA = 15$

Parallel chords in a circle have equal arcs.

Example: In $\odot O$, $\overline{AB} \parallel \overline{CD}$. If $\overset{\frown}{AC} = 38°$ and $\overset{\frown}{BD} = (3x - 7)°$, find x.

Solution: Because $\overline{AB} \parallel \overline{CD}$, $\overset{\frown}{AC} \cong \overset{\frown}{BD}$.
$38 = 3x - 7$ so $x = 15°$.

1. In circle $\odot O$, if $\overset{\frown}{AC} \cong \overset{\frown}{BD}$ then
   1) $\overline{AB} \cong \overline{CD}$
   2) $\overline{AB} \parallel \overline{CD}$
   3) $\overline{AB} \cong \overline{CD}$
   4) $\overline{CD} > \overline{AB}$

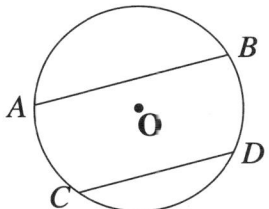

3. In circle $\odot O$, if $\overset{\frown}{AC} \cong \overset{\frown}{BD}$, $\overset{\frown}{AC} = (5x-8)°$ and $\overset{\frown}{BD} = 72°$, find x in degrees.

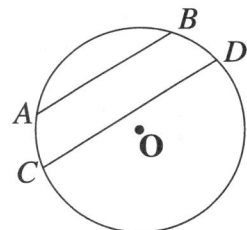

2. In circle $\odot O$, $\overline{OT} \perp \overline{PT}$, $\overline{OT} = x$, $\overline{PT} = x + 14$ and $\overline{PDO} = x + 16$. Find $\overline{BOP}$.

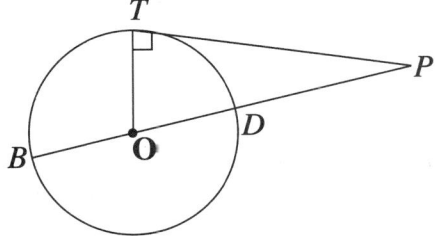

4. In circle $\odot O$, $\overline{ON} \perp \overline{NB}$, $\overline{ON} = x$, $BN = 12$ and $\overline{OB} = x + 6$. Find $\overline{ON}$.

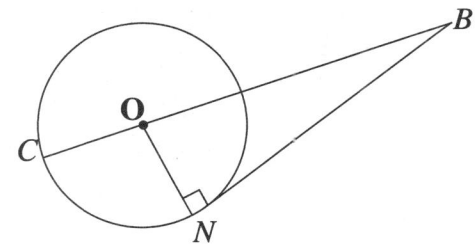

Chapter 4

MEASUREMENT IN CIRCLES

Common External and Internal Tangents

> **REMEMBER**
>
> Common external tangents are equal:
>
> a) two non-intersecting circles $t_1 = t_2$
> b) two tangent circles $t_1 = t_2$
> c) two overlapping circles $t_1 = t_2$
>
>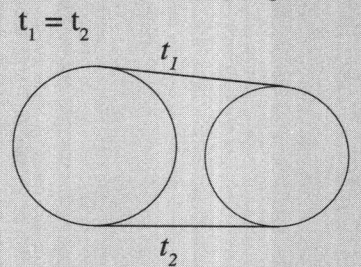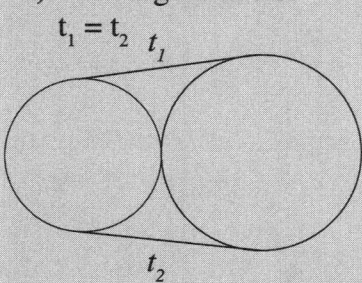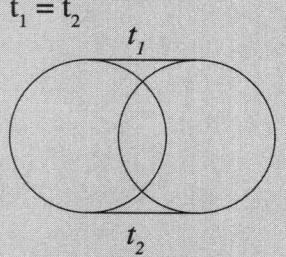
>
> Common internal tangents are equal:
>
> a) $t_1 = t_2$
> b) $t_1 = t_2$

1. Tangent t_1 and t_2 are drawn to $\odot O_1$ and $\odot O_2$. If $t_1 = 8x + 4$ and $t_2 = 6x + 68$, find t_1.

   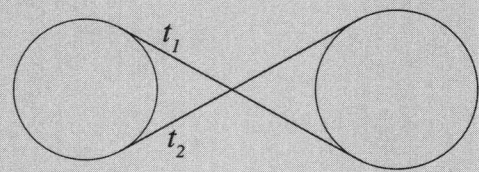

   _____

2. If an external tangent AB is draw to intersecting circles $\odot O_1$ and $\odot O_2$, find the length of tangent CD if $O_1A = 6$, $O_2B = 10$, $PA = 8$ and $PB = x + 8$.

   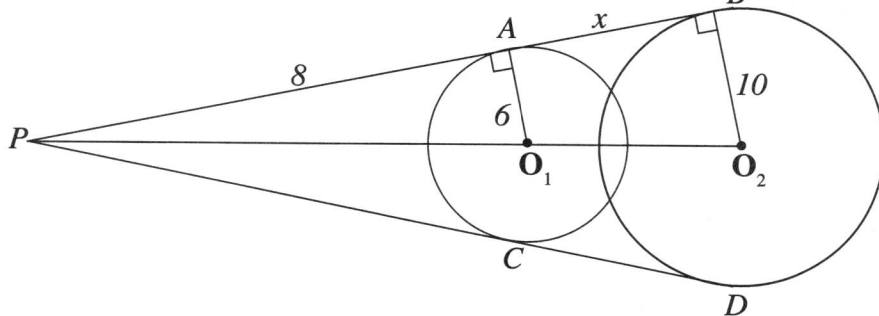

   _____

Chapter 4

MEASUREMENT IN CIRCLES

Common External and Internal Tangents (continued)

3. $\odot O_1$ and $\odot O_2$ are tangent at T, $O_1T = x$, $O_2T = x + 5$, and $O_1B = 17$, $O_1O_2 = 35$. Find AB.

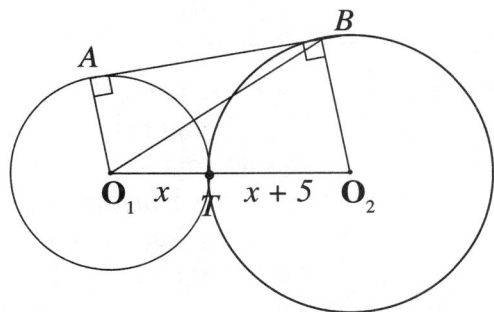

4. Two non-intersecting circles are drawn with internal tangents $\overline{AB}$ and $\overline{CD}$. If $O_1A = 15$, $O_2D = 8$, $O_1P = 17$ and $O_2P = 10$, find the length of one internal tangent.

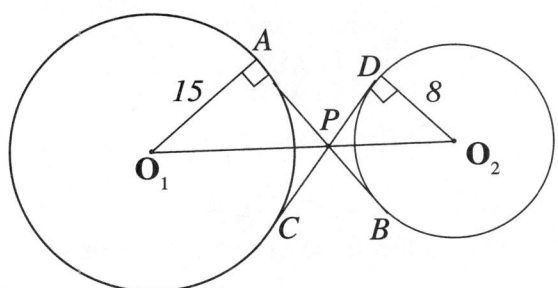

5. In the accompanying figure, $\overline{AB}$ is tangent to $\odot O$ at point D, $\overline{BC}$ is tangent to $\odot O$ at point E and $\overline{AC}$ is tangent to $\odot O$ at point F. Find the perimeter of $\triangle ABC$.

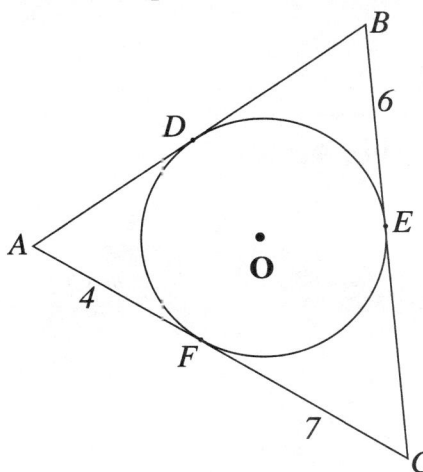

Chapter 5

COORDINATE GEOMETRY

Find the slope of a line that is perpendicular to the given equation of a line.

> **REMEMBER**
> The slopes of two perpendicular lines are negative reciprocals of each other.

1. a Write the equation of a line that is ⊥ to $y = 2x - 8$ and has a y–intercept of 5.
    b Write the equation of a line that is ⊥ to $7x + 5y = 35$ and has a y–intercept of -1.
    c Write the equation of a line that has the same y–intercept as $y = x + 2$ and ⊥ to $y = x + 2$.
    d Write the equation of a line that is perpendicular to $y = x$ and passes through the origin.
    e Write the equation of a line that is ⊥ to the x-axis and 3 units to the left of the y–axis.
    f Write the equation of a line that is perpendicular to the x-axis and 5 units to the right of the y–axis.
    g Write the equation of a line that is perpendicular to $y = \frac{1}{2}x + 6$ and has the same y-intercept as the equation of the line $y = -9x - 7$
    h Write the equation of a line that passes through the point (3,8) and is ⊥ to the y-axis.

    a

    b

    c

    d

    e

    f

    g

    h

2. If the slope of $\overline{RT}$ is $\frac{5}{9}$, and $\overline{RT} \perp \overline{QW}$, what is the slope of $\overline{QW}$? Explain.

3. What is the slope of the line that is perpendicular to the line that passes through the points (8, –6) and (–2, 4)? Show your work.

Chapter 5

COORDINATE GEOMETRY

Determine whether two lines are parallel, perpendicular, or neither, given their equations.

> **REMEMBER**
>
> The slopes of two perpendicular lines are negative reciprocals of each other.
> The slopes of two parallel lines are equal.
> The slope of $y = x$ is 1.   The slope of $y = -x$ is -1.
> The slope of a horizontal line, $y = k$, is 0.
> A vertical line, $x = k$, has *no slope*.

1. Show why the equations of the two lines given are parallel, perpendicular or neither by finding their slopes. Explain the answers.

a	$y = 3x - 6$ and $y = -3x - 6$	h	$y = 4$ and $y = -2$
b	$y = -5x - 8$ and $y = -5x + 8$	i	$y = -6$ and $x = 0$
c	$y = 4x + 5$ and $y = 2x + 5$	j	$y = 2x + 8$ and $y = -2$
d	$y = -8x + 2$ and $y = -7x - 2$	k	$x = 9$ and $x = -6$
e	$y = \frac{1}{2}x + 9$ and $y = 0.5x - 9$	l	$y = x$ and $y = -x$
f	$y = \frac{1}{4}x - 6$ and $y = 4x + 6$	m	$x - 3y = 0$ and $3x + y = 0$
g	$y = \frac{1}{3}x + 3$ and $y = -3x + 3$	n	$x + 2y + 2 = 0$ and $4x - 2y + 2 = 0$

Chapter 5

COORDINATE GEOMETRY

Find the equation of a line given a point on the line and the equation of a line perpendicuar to the given line

> **REMEMBER**
>
> The slopes of two perpendicular lines are negative reciprocals of each other.
> **Example:** Find the equation of the line perpendicular to $3x + 4y = 12$ and Passing through the point $(-1, 3)$.
> **Solution:** $3x + 4y = 12$
> $4y = -3x + 12$
> $y = -\frac{3}{4}x + 3$ where the slope, m, $= -\frac{3}{4}$
> The slope of a line perpendicular is the negative reciprocal.
> $y_2 - y_1 = m(x_2 - x_1)$
> $y - 3 = \frac{4}{3}(x + 1)$
> $3y - 9 = 4x + 4$ or $4x - 3y = -13$ Answer

1. Find the equation of a line passing through the point $(0, 0)$ and perpendicular to the line $y = 6x - 3$.

2. Find the equation of a line passing through the point $(0, 4)$ and perpendicular to the line $y = -5x - 8$.

3. Find the equation of a line passing through the point $(-5, 0)$ and perpendicular to the line $y = 0.5x - 1$.

4. Find the equation of a line passing through the point $(5, 3)$ and perpendicular to the line $y = -3$.

5. Find the equation of a line passing through the point $(-1, -8)$ and perpendicular to the line $2x + y - 1 = 0$.

6. Find the equation of a line passing through the point $(10, -6)$ and perpendicular to the line $x = 6$.

7. Find the equation of a line passing through the point $(-9, -3)$ and perpendicular to the line $x + 2y + 2 = 0$

Chapter 5

COORDINATE GEOMETRY

Find the equation of a line given a point on the line and the equation of a line parallel to the given line.

> **REMEMBER**
>
> The slopes of two parallel lines are equal to each other.
>
> **Example:** Find the equation of the line parallel to $3x + 4y = 12$ and passing through the point $(-1, 3)$.
>
> **Solution:** $3x + 4y = 12$
>
> $4y = -3x + 12$
>
> $y = \dfrac{-3}{4}x + 3$ where the slope, $m, = \dfrac{-3}{4}$
>
> The slopes of parallel lines are equal to each other.
>
> $y_2 - y_1 = m(x_2 - x_1)$
>
> $y - 3 = \dfrac{-3}{4}(x + 1)$
>
> $4y - 12 = -3x - 3$ or $-3x - 4y = -9$ Answer

1. Find the equation of a line passing through the point $(0, 0)$ and parallel to the line $y = 6x - 3$.

2. Find the equation of a line passing through the point $(0, 4)$ and parallel to the line $y = -5x - 8$.

3. Find the equation of a line passing through the point $(-5, 0)$ and parallel to the line $y = 0.5x - 1$.

4. Find the equation of a line passing through the point $(5, 3)$ and parallel to the line $y = -3$.

5. Find the equation of a line passing through the point $(-1, -8)$ and parallel to the line $2x + y - 1 = 0$.

6. Find the equation of a line passing through the point $(10, -6)$ and parallel to the line $x = 6$.

7. Find the equation of a line passing through the point $(-9, -3)$ and parallel to the line $x + 2y + 2 = 0$.

Chapter 5

COORDINATE GEOMETRY

Find the midpoint of a line segment, given its endpoints

REMEMBER

$$\text{midpoint}, (x, y) = \left(\frac{x_1 + x_2}{2}, \frac{y_1 + y_2}{2} \right)$$

Example: The diameter of a circle has endpoints (3, 2) and (9, 8). Find the center of the circle.

Solution: $(x, y) = \left(\frac{3+9}{2}, \frac{2+8}{2} \right) = (6, 5)$ Answer

1. Triangle TAR has coordinates T(5,–2), A(3,1), and R(–3,2). What are the coordinates of the point M where AM is the median to side TR?

2. Parallelogram QUAD has coordinates Q(0,0), U(4,8), A(10,4), and D(6,–4). What are the coordinates of the point of intersecton of the diagonals?

3. The endpoint of a segment is (–8,4). The midpoint is (–1,–2). Find the other endpoint.

4. Circle O has a center (3,–5) and a diameter $\overline{AB}$. The coordinates of B are (–3,6). What are the coordinates of A?

Chapter 5

COORDINATE GEOMETRY

Find the slope of a line that is perpendicular to the given equation of a line.

> **REMEMBER**
>
> Length of a segment = $\sqrt{(x_2-x_1)^2 + (y_2-y_1)^2}$
>
> **Example 1:** The diameter of a circle has endpoints (3,2) and (9,8). Find the length of the diameter of the circle in radical form.
>
> **Solution 1:** Length of a segment = $\sqrt{(9-3)^2 + (8-2)^2}$
> $= \sqrt{(6)^2 + (6)^2} = \sqrt{72} = 6\sqrt{2}$ Answer
>
> **Example 2:** Find the distance between points (-3, -2) and (3, 4).
>
> **Solution 2:** Plot the points.
> Add up the horizontal and vertical boxes.
> Use the Pythagorean theorem.
>
> Distance: $= \sqrt{(-3-3)^2 + (-2-4)^2}$
> $= \sqrt{(-6)^2 + (-6)^2} = 6\sqrt{2}$ Answer
>
>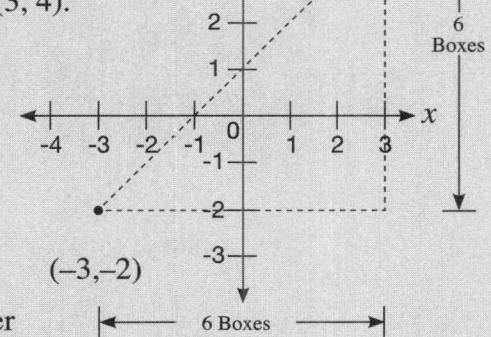

1. What is the length of the line segment that joins points (6,–2) and (9,4)?

   (1) $3\sqrt{5}$ (3) 3
   (2) $\sqrt{5}$ (4) $5\sqrt{3}$

2. What is the distance between points A (–4,7) and B (–7,5)?

   (1) $\sqrt{5}$ (3) $\sqrt{13}$
   (2) 13   (4) $\sqrt{45}$

3. Find in *radical form,* the distance between points (–5,6) and (4,–2).

4. Find to the *nearest tenth,* the distance between points (–9,–6) and (6,3).

77

Chapter 5

COORDINATE GEOMETRY

Find the equation of a line that is the perpendicular bisector of a line segment, given the endpoints of the line segment.

> **REMEMBER**
>
> **Example:** Find the equation of a line that is the perpendicular bisector of a line segment, given the endpoints of the line segment are $(-5, -3)$ and $(7, 5)$.
>
> **Solution:** Find the midpoint of the line segment since it is being bisected.
>
> $$\text{midpoint}, (x, y) = \left(\frac{x_1 + x_2}{2}, \frac{y_1 + y_2}{2}\right) = (1, 1)$$
>
> $$\text{slope, m, of line segment} = \frac{y_2 - y_1}{x_2 - x_1} = \frac{8}{12} = \frac{2}{3}$$
>
> $$\text{slope of a perpendicular line} = -\frac{3}{2}$$
>
> Equation of the perpendicular bisector:
>
> $$y_2 - y_1 = \text{slope}\,(x_2 - x_1)$$
>
> $$y - 1 = -\frac{3}{2}(x - 1) \qquad \text{using midpoint } (1, 1)$$
>
> $$2y - 2 = -3(x - 1)$$
>
> $$3x + 2y = 5 \qquad \text{Answer}$$

1. Find the equation of a line that is the perpendicular bisector of a line segment, given the endpoints of the line segment as specified below.

a	$(0, 0)$ and $(10, 0)$	d	$(20, 12)$ and $(20, -6)$
b	$(5, 0)$ and $(-5, 0)$	e	$(2, 15)$ and $(10, 27)$
c	$(-6, 8)$ and $(-6, -8)$	f	$(14, 1)$ and $(-10, 9)$

Chapter 5

COORDINATE GEOMETRY

Properties of triangles in the coordinate plane using the distance, midpoint, and slope formulas

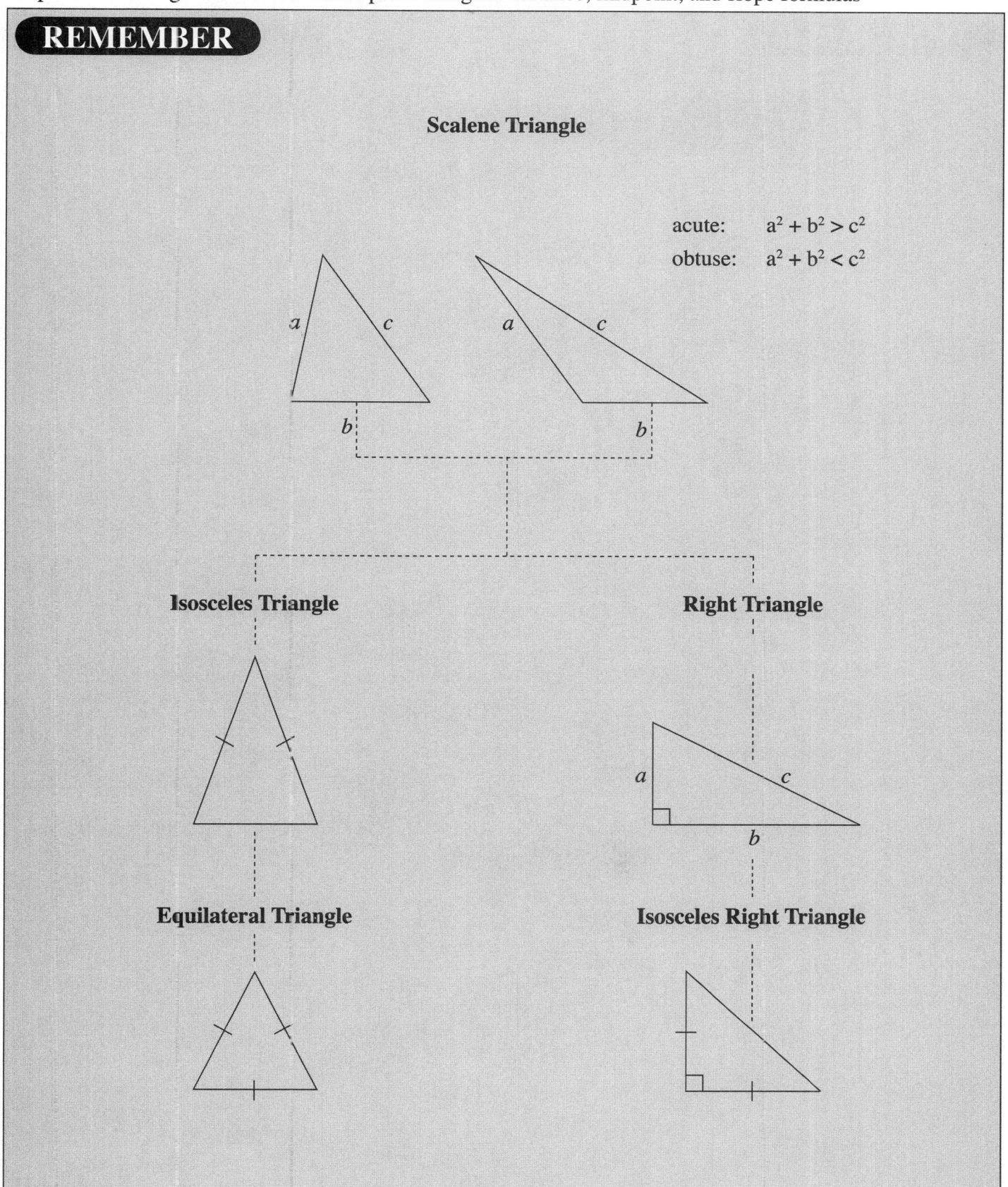

Chapter 5

COORDINATE GEOMETRY

Properties of quadrilaterals in the coordinate plane using the distance, midpoint, and slope formulas

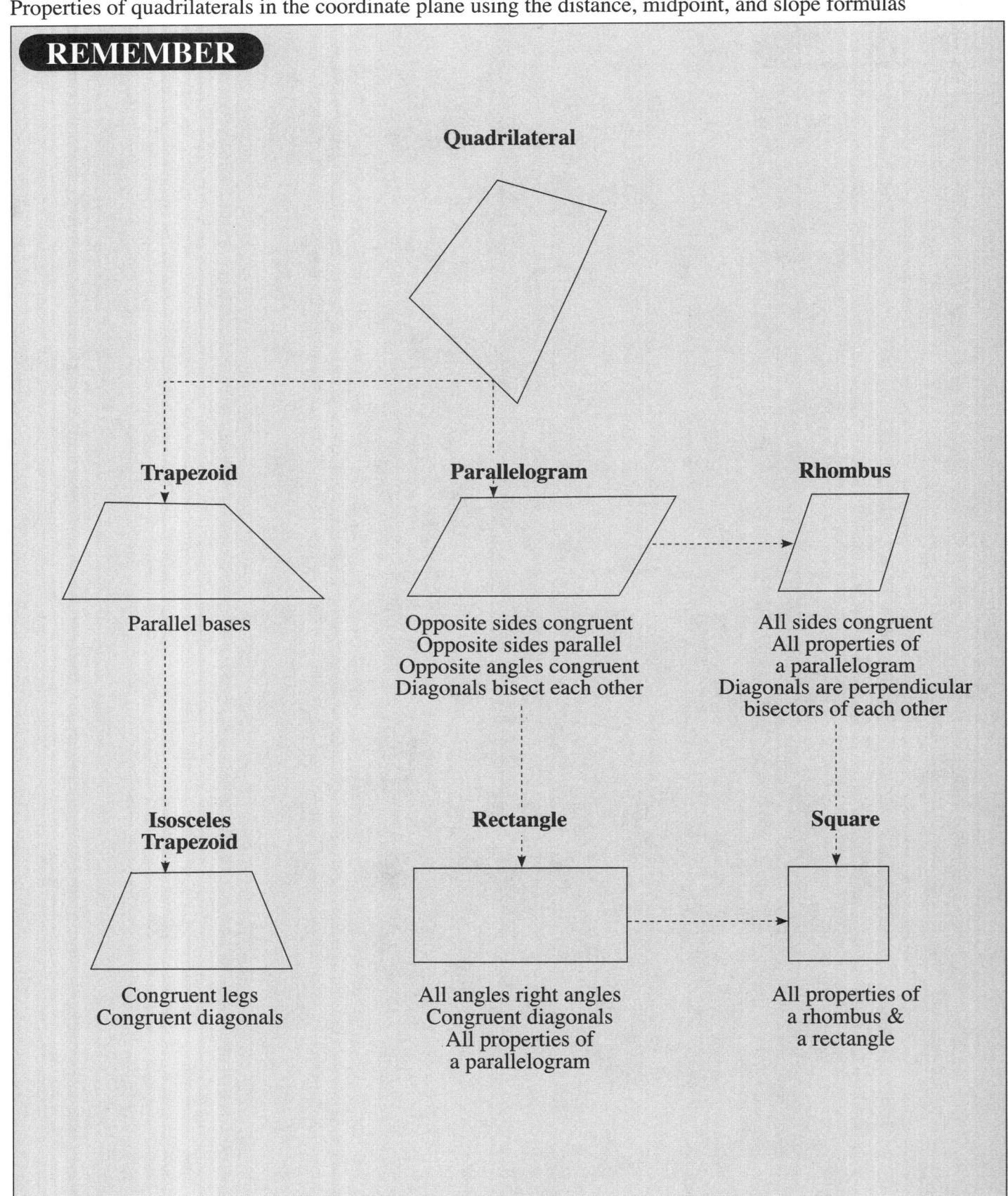

Chapter 5

COORDINATE GEOMETRY

Properties of triangles and quadrilaterals in the coordinate plane using the distance, midpoint, and slope formulas

> **REMEMBER**
>
> length of a segment = $\sqrt{(x_2 - x_1)^2 + (y_2 - y_1)^2}$
>
> midpoint, (x, y) of a segment = $\left(\dfrac{x_1 + x_2}{2}, \dfrac{y_1 + y_2}{2}\right)$
>
> slope of line segment: $m = \dfrac{y_2 - y_1}{x_2 - x_1}$
>
> or $\quad y_2 - y_1 = m(x_2 - x_1)$

1. Triangle AFN has vertices A(–7, 6), F(–1, 6) and N(–4, 2). Prove triangle AFN is an isosceles triangle but *not* equilateral.

2. Triangle MEP has vertices M(11, 14), E(6, 4) and P(3, 8). Show triangle MEP is a right triangle but *not* isosceles.

3. The vertices of triangle RST are R(6, 10), S(8, 2) and T(0, 2).

   Find the coordinates of point X, the midpoint of segment RS.

   Find the coordinates of point Y, the midpoint of segment RT.

   Find the coordinates of point Z, the midpoint of segment ST.

   Find the length of the medians TX, SY and RZ.

   What is the point of intersection of the medians called?

Chapter 5
COORDINATE GEOMETRY

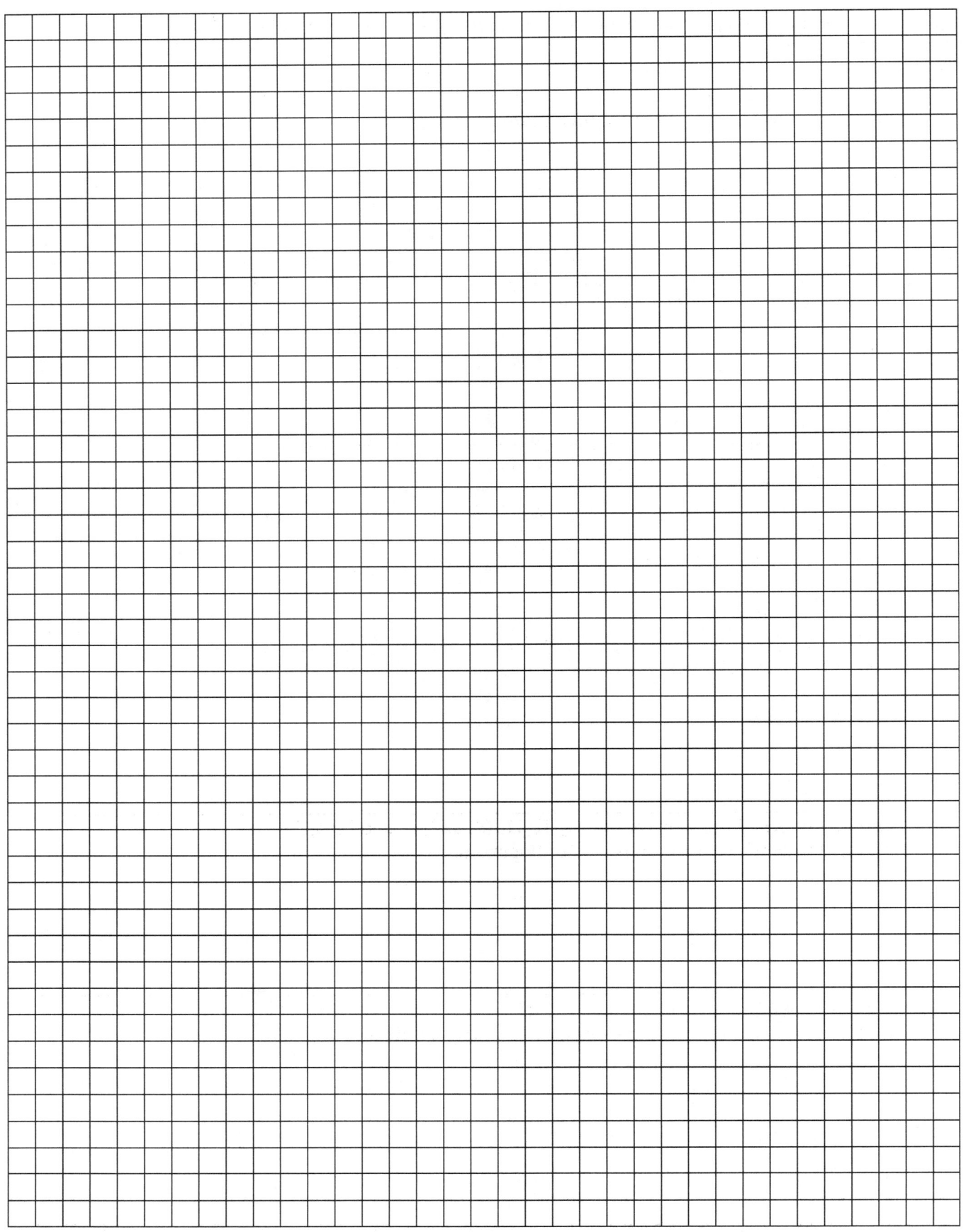

Chapter 5

COORDINATE GEOMETRY

Properties of triangles and quadrilaterals in the coordinate plane using the distance, midpoint, and slope formulas (continued)

1. Quadrilateral WXZY has vertices W(2, 2), X(5, 2), Y(3, –2) and Z(6, –2).
   Prove quadrilateral WXZY is a parallelogram.

2. Quadrilateral ABCD has vertices A(2, 3), B(10, 3), C(10, –1) and D(2, –1).
   Prove quadrilateral ABCD is a rectangle.

3. Quadrilateral QRST has vertics Q(6, 7), R(11, 7), S(8, 3) and T(3, 3).
   Prove quadrilateral QRST is a rhombus.

4. Quadrilateral EFGH has vertices E(–7, 0), F(–2, 0), G(–2, –5) and H(–7, –5).
   Prove quadrilateral EFGH is a square.

5. Quadrilateral JKLM has vertices J(4, 7), K(11, 0), L(7, 0) and M(4, 3).
   Prove quadrilateral JKLM is an isosceles trapezoid.

6. Quadrilateral ABCD has vertices A(7, 8), B(12, 8), C(4, 4) and D(9, 4).
   Prove that the diagonals of the quadrilateral are perpendicular bisectors of each other.

Chapter 5
COORDINATE GEOMETRY

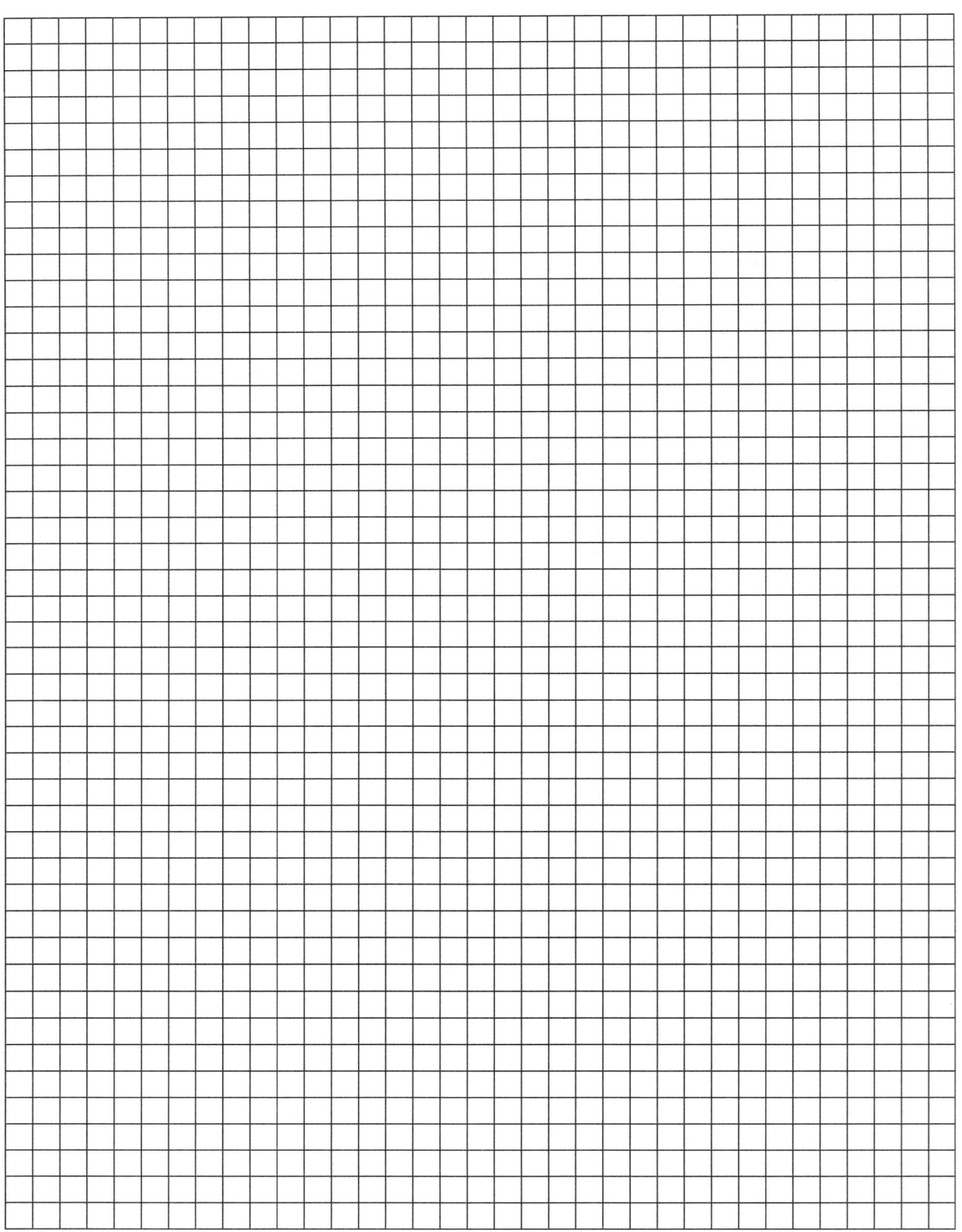

Chapter 5

COORDINATE GEOMETRY

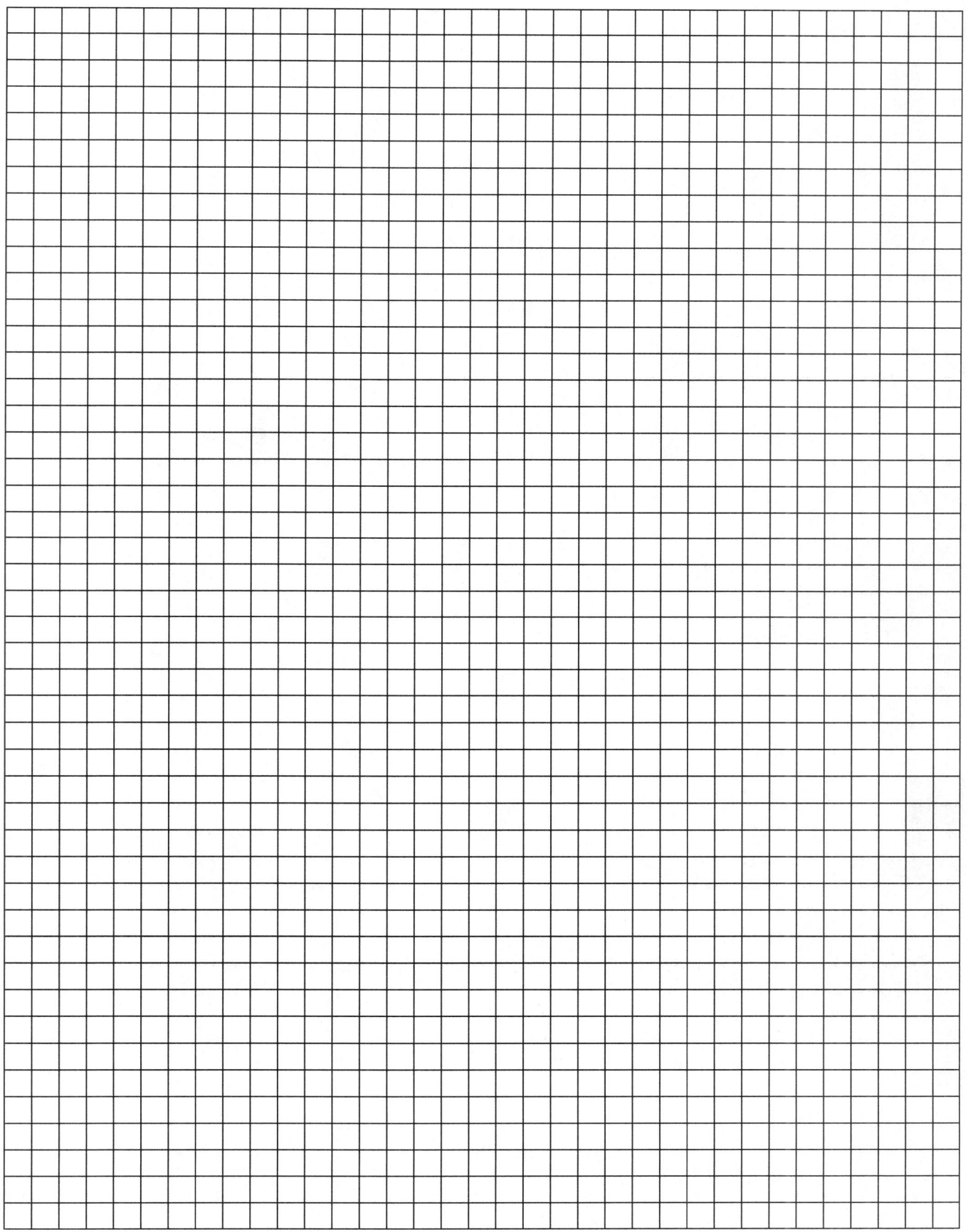

Chapter 5
COORDINATE GEOMETRY

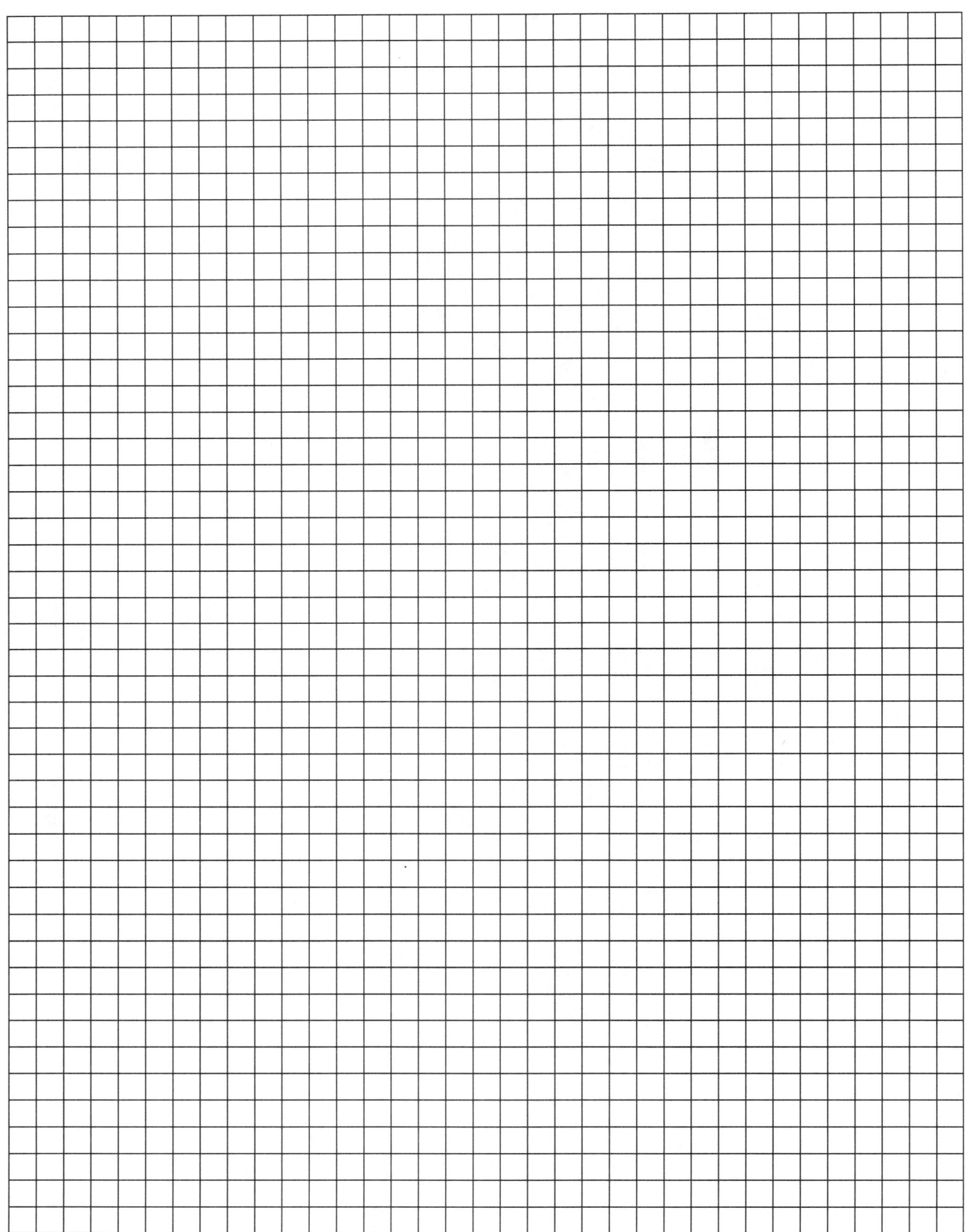

Chapter 5

COORDINATE GEOMETRY

Solve systems of equations involving one linear equation and one quadratic equation graphically

> **REMEMBER**
>
> **Example 1:** Determine where the graphs of $y = x^2 - 4x + 9$ and $y = 2x + 1$ intersect by graphing each function on the same coordinate axis system.
>
> **Solution 1:**
>
>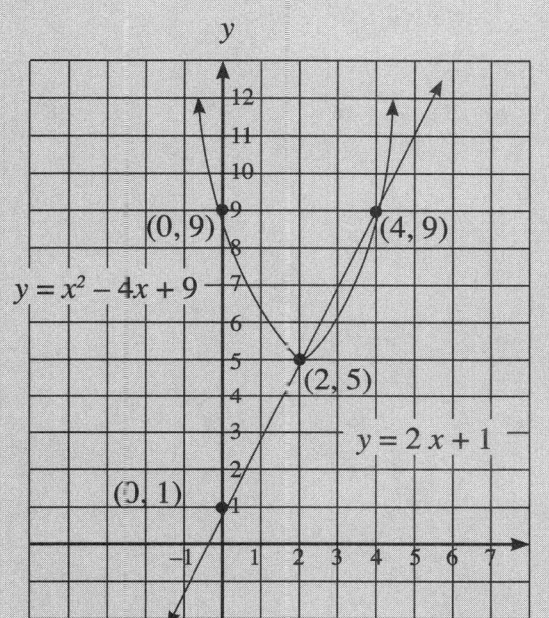
>
> The graphs intersect at $\{(2, 5), (4, 9)\}$.
>
> **Example 2:** Determine the points of intersection for the graphs of $x^2 + (y + 1)^2 = 25$ and $y = 3$ by graphing each equation on the same coordinate axis system.
>
> **Solution 3:**
>
>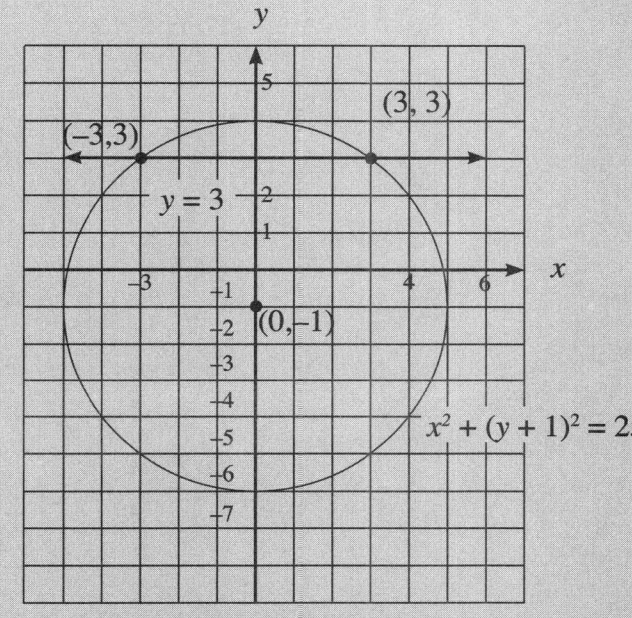
>
> The graphs intersect at $\{(-3, 3), (3, 3)\}$.

1. Graphically determine the points of intersection for the fuctions $y = x^2 - x - 1$ and $y = 2x + 3$.

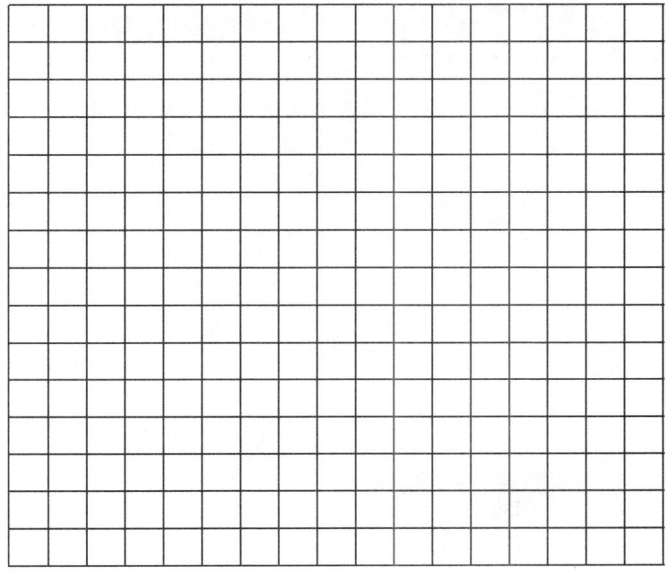

87

Chapter 5

COORDINATE GEOMETRY

Write the equation of a circle, given its center and radius or given the endpoints of a diameter

REMEMBER

Example 1: Write the equation of a circle having a center of (2, –4) and a radius of 7.

Solution 1: $(x - h)^2 + (y - k)^2 = r^2$

$(x - 2)^2 + (y + 4)^2 = 49$ Answer

Example 2: Write the equation of a circle with the diameter having the endpoints (5, –1) and (–5, –1).

Solution 2: Length of diameter = $\sqrt{(x_2 - x_1)^2 + (y_2 - y_1)^2}$

Length of diameter = $\sqrt{(5 - -5)^2 + (-1 - -1)^2}$

Length of diameter = $\sqrt{(10)^2 + (0)^2}$

Length of diameter = 10, $r = 5$, $r^2 = 25$

Midpoint, $(x, y) = \left(\dfrac{x_1 + x_2}{2}, \dfrac{y_1 + y_2}{2}\right)$

Midpoint, $(x, y) = \left(\dfrac{5 + -5}{2}, \dfrac{-1 + -1}{2}\right)$

Midpoint, $(x, y) = (0, -1)$ $\therefore h = 0$ and $k = -1$

$(x - h)^2 + (y - k)^2 = r^2$

$x^2 + (y + 1)^2 = 25$ Answer

---

1. Determine the equation of a circle with center (–1, 7) and a diameter length of 16.

2. Write the equation of a circle whose diameter's endpoints are the points (3, –8) and (–1, –5).

Chapter 5

COORDINATE GEOMETRY

Write the equation of a circle, given its graph (The center is an ordered pair of integers and the radius is an integer

3. Write the equation of a circle having the graphs shown below:

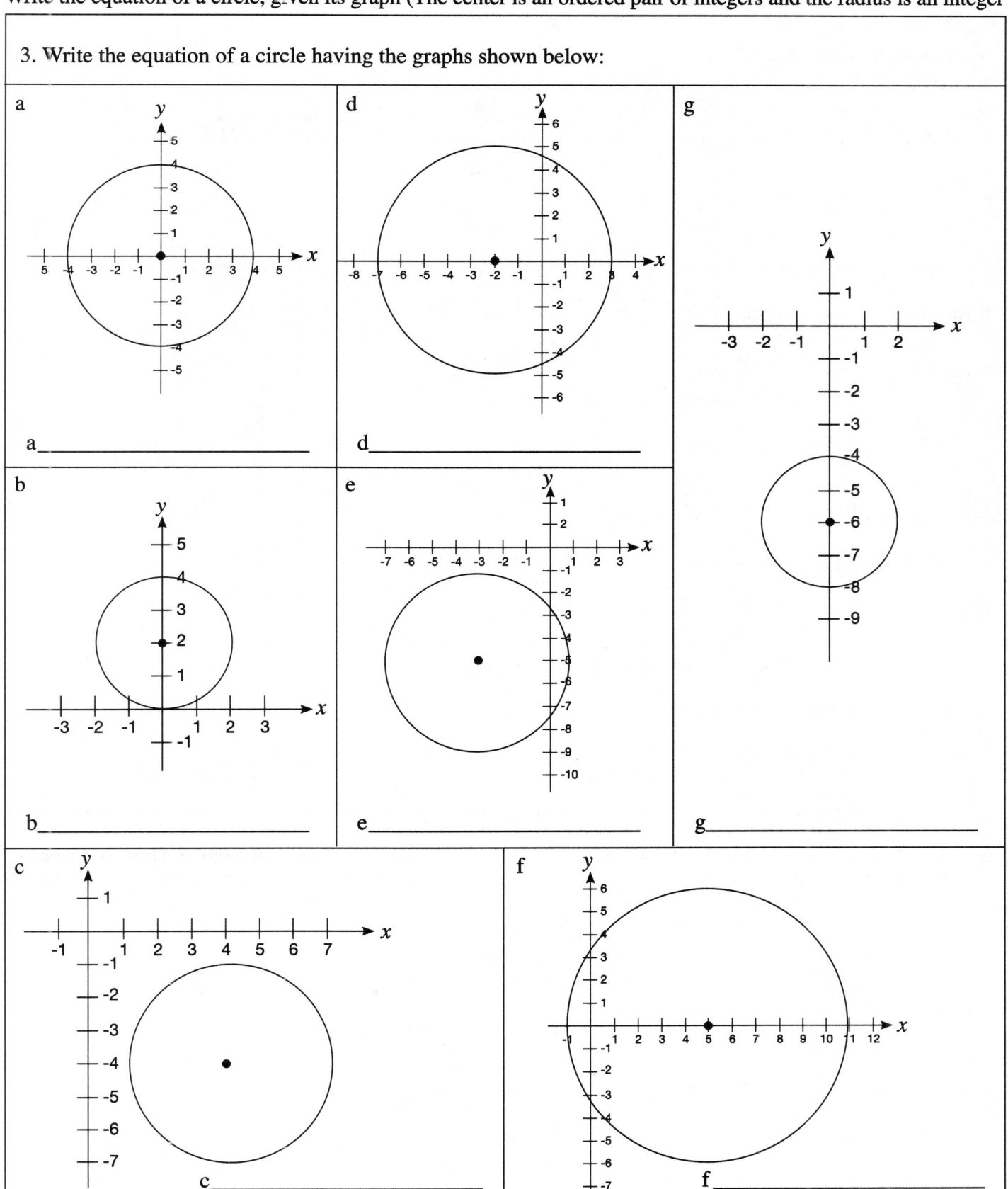

Chapter 5

COORDINATE GEOMETRY

Find the center and radius of a circle, given the equation of the circle in center radius form $(x - h)^2 + (y - k)^2 = r^2$

REMEMBER

Example:
Describe the circle whose equation is given by:

		Solutions:		
			center	radius
a	$(x - h)^2 + (y - k)^2 = r^2$	a	(h, k)	r
b	$(x - 4)^2 + (y - 2)^2 = 16$	b	$(4, 2)$	4
c	$(x + 1)^2 + (y + 3)^2 = 9$	c	$(-1, -3)$	3
d	$(x - 5)^2 + (y + 7)^2 = 25$	d	$(5, -7)$	5
e	$x^2 + (y + 6)^2 = 81$	e	$(0, -6)$	9

Example: The equation of a sphere with center (h, j, k) and radius r is:
$$(x - h)^2 + (y - j)^2 + (z - k)^2 = r^2$$

Determine the center and radius of a sphere if its equation is:
$$(x - 1)^2 + (y + 3)^2 + (z - 2)^2 = 36$$

Solution: center is $(1, -3, 2)$ with a radius of 6   Answer

---

1. Find the center and radius of a circle with the given equation:

    a $(x + 12)^2 + (y + 3)^2 = 36$

    b $(x - 8)^2 + (y - 6)^2 = 25$

    c $(x - 9)^2 + (y + 3)^2 = 100$

    d $(x + 12)^2 + (y - 2)^2 = 49$

    e $x^2 + (y - 1)^2 = 10$

    f $(x - 6)^2 + (y - 8)^2 = 15$

    g $(x - 2)^2 + (y + 6)^2 = 4$

    h $(x - 18)^2 + (y - 26)^2 = 53$

    i $(x + 16)^2 + (y - 26)^2 = 144$

    j $(x - 14)^2 + y^2 = 121$

2. Find the center and radius of a sphere with the given equation:

    a $(x - 3)^2 + (y + 6)^2 + (z - 7)^2 = 49$

    b $(x + 9)^2 + y^2 + (z + 12)^2 = 10^2$

Chapter 5

COORDINATE GEOMETRY

Graph circles of the form $(x - h)^2 + (y - k)^2 = r^2$

REMEMBER

Example: Graph the circles whose equation is given by:

a) $(x + 3)^2 + (y - 5)^2 = 4$

Solution: center of circle is $(-3, 5)$

radius is $\sqrt{4} = 2$

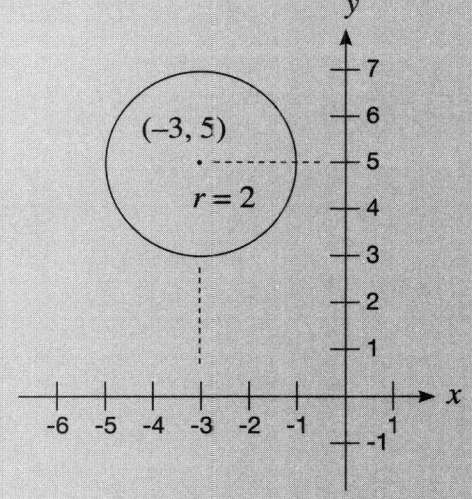

b) $(x - 4)^2 + y^2 = 9$

Solution: center of circle is $(4, 0)$

radius is $\sqrt{9} = 3$

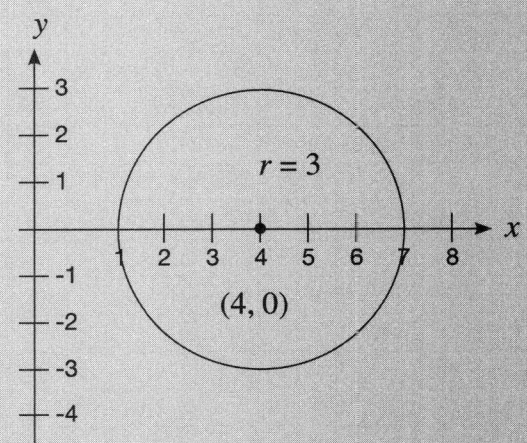

Graph the equations of the circles and label the center and radius for each:

1. $(x - 4)^2 + (y - 2)^2 = 16$
2. $(x - 5)^2 + (y + 7)^2 = 25$
3. $(x - 2)^2 + y^2 = 36$
4. $(x + 1)^2 + (y + 3)^2 = 9$
5. $x^2 + (y + 6)^2 = 81$
6. $(x + 3)^2 + (y - 5)^2 = 49$

Chapter 5
COORDINATE GEOMETRY

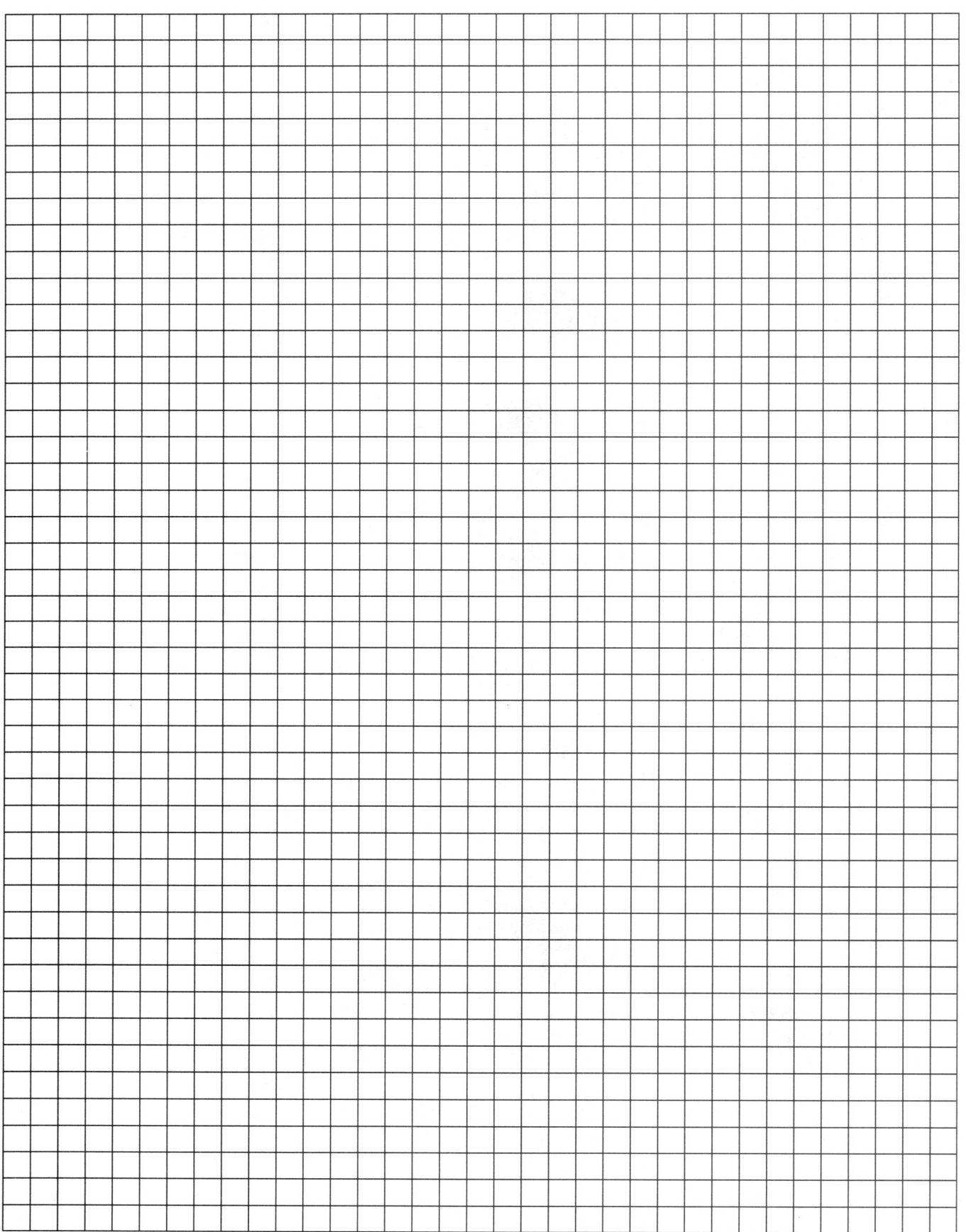

Chapter 5
COORDINATE GEOMETRY

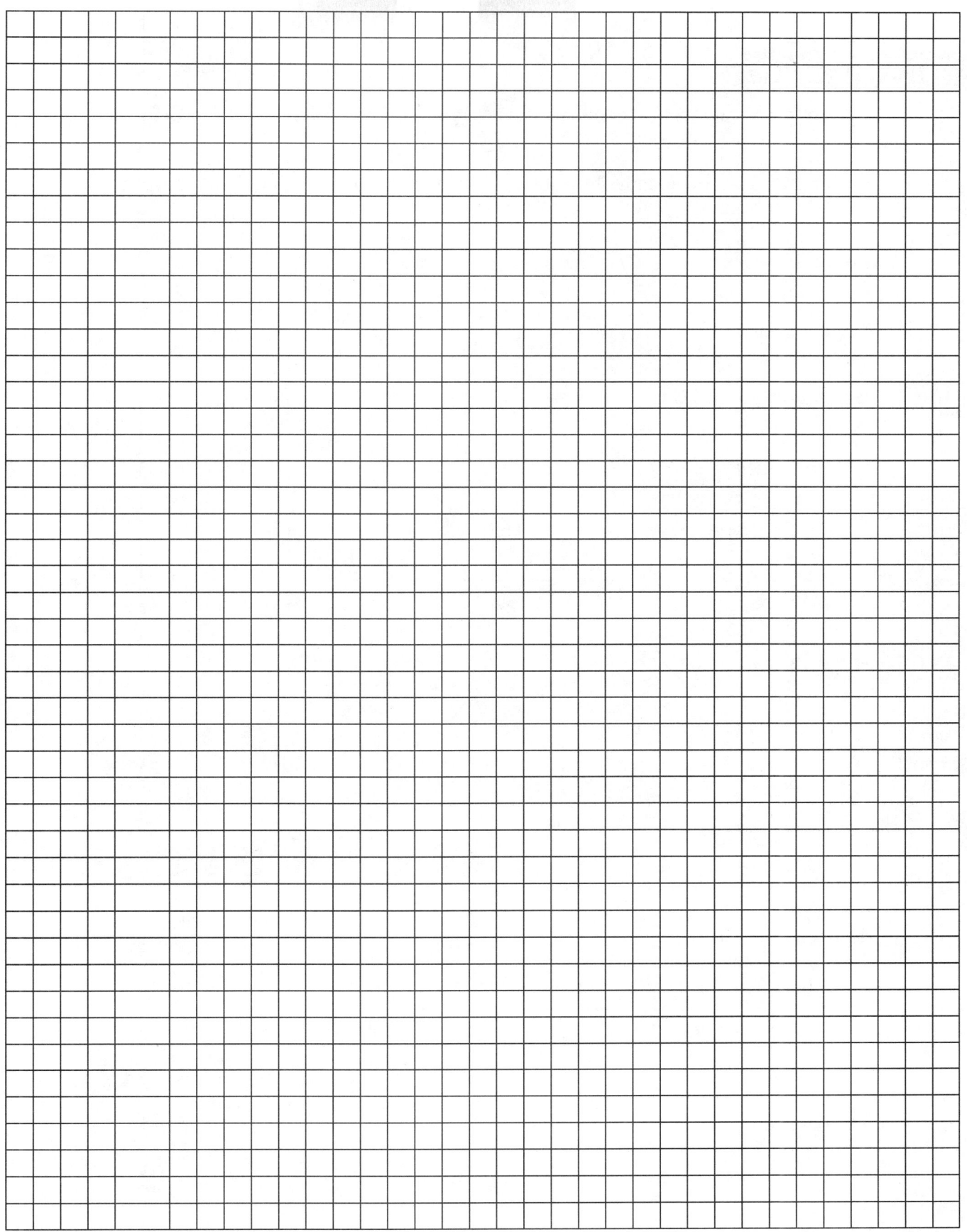

Chapter 5

COORDINATE GEOMETRY

Given the coordinates of a rectangle, prove the diagonals are congruent using the distance formula.

> **REMEMBER**
>
> **Example:** A rectangle has the coordinates A(0, 8), B(10, 8), C(10, 0) and D(0, 0). Prove the diagonals are congruent using the distance formula.
>
> **Solution:** Find the length of diagonal AC and diagonal BD. They should have equal values in length.
>
> length of segment AC = $\sqrt{(x_2 - x_1)^2 + (y_2 - y_1)^2}$
>
> length of segment AC = $\sqrt{(10 - 0)^2 + (0 - 8)^2}$
>
> length of segment AC = $\sqrt{100 + 64}$
>
> length of segment AC = $\sqrt{164}$
>
> length of segment BD = $\sqrt{(x_2 - x_1)^2 + (y_2 - y_1)^2}$
>
> length of segment BD = $\sqrt{(10 - 0)^2 + (8 - 0)^2}$
>
> length of segment BD = $\sqrt{100 + 64}$
>
> length of segment BD = $\sqrt{164}$
>
> AC = BD = $\sqrt{164}$ **Answer**   The diagonals are equal in measure.

1. Given the coordinates of a rectangle, prove the diagonals are congruent using the distance formula:

   a   A(–12, –2), B(–4, –2), C(–4, –10), and D(–12, –10).

   b   E(0, 10), F(8, 6), G(5, 0), and H(–3, 4).

Chapter 5
COORDINATE GEOMETRY

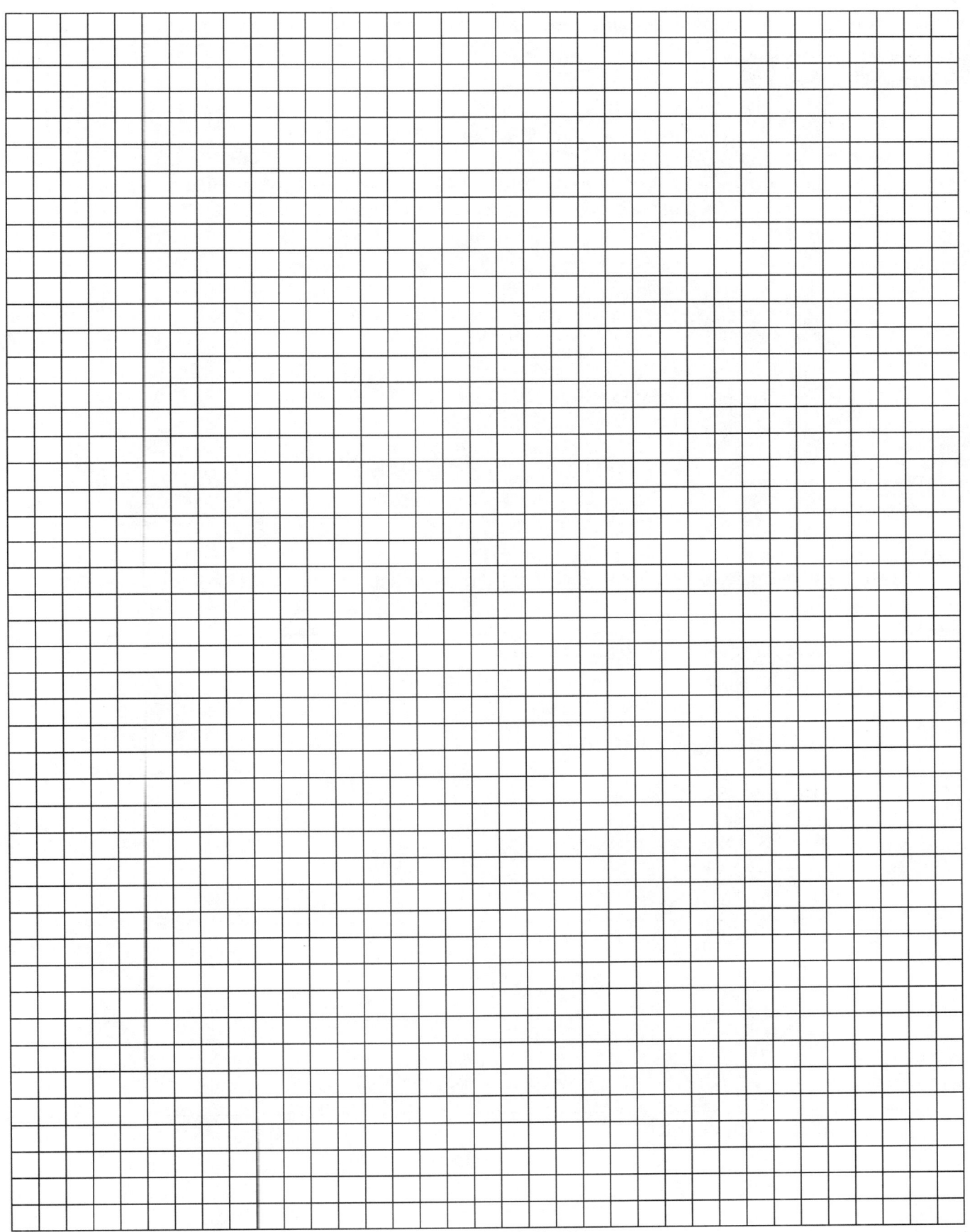

Chapter 5

COORDINATE GEOMETRY

Given the coordinates of a triangle, determine the coordinates of the centroid of the triangle.

> **REMEMBER**
>
> **Example:** Given triangle ABC with vertices A(2, 8), B(8, 12) and C(8, 4), determine the coordinates of the centroid of the triangle.
>
> **Solution:** The coordinates of the centroid of any triangle are the means of the domain and the range of the vertices.
>
> $$Q(x, y) = \left(\frac{x_a + x_b + x_c}{3}, \frac{y_a + y_b + y_c}{3} \right)$$
>
> $$Q(x, y) = \left(\frac{(2 + 8 + 8)}{3}, \frac{(8 + 12 + 4)}{3} \right) = (6, 8) \quad \text{Answer}$$

1. Given the coordinates of a triangle, determine the coordinates of the centroid of each triangle

   a   D(0, 0), E(3, 15), and F(12, 0)

   b   G(1, 0), H(1, 6), and I(10, 0)

   c   J(0, 0), K(15, 0), and L(−3, 3)

   d   M(3, 15), N(12, 0), and P(−3, −3)

   e   Q(−2, 0), R(−4, −3), and S(−12, −6)

   f   T(4, 5), U(6, 1), and V(8, 9)

Chapter 5

COORDINATE GEOMETRY

Given the coordinates of a triangle, find the lengths of the segments of the medians

> **REMEMBER**
>
> **Example:** Given triangle ABC with the coordinates A(4, 8), B(12, 2), and C(4, 2), find the lengths of the segments of the medians.
>
> **Solution:** 1. Find the midpoints of each side of the triangle. This will give a second point on each median.
> 2. Then use the distance formula to find the length of each median.
>
> Let the medians be AN, BQ and CP.
>
> N, midpoint of BC = $\left(\dfrac{4+12}{2}, \dfrac{2+2}{2}\right)$ = (8, 2)
>
> P, midpoint of AB = $\left(\dfrac{4+12}{2}, \dfrac{8+2}{2}\right)$ = (8, 5)
>
> Q, midpoint of AC = $\left(\dfrac{4+4}{2}, \dfrac{8+2}{2}\right)$ = (4, 5)
>
> median, AN = $\sqrt{(4-8)^2 + (8-2)^2}$ = $\sqrt{16+36}$
>
> median, AN = $\sqrt{52}$ = $2\sqrt{13}$
>
> median, BQ = $\sqrt{73}$ median, CP = 5

1. Find the lengths of the segments of the medians for the triangles with the given vertices:

   a   D(3, 8), E(9, 8), and F(17, 10)

   b   G(0, –3), H(–2, –9), and J(8–9)

   c   K(3,–1), L(9, 5), and M(–3, 2)

Chapter 5
COORDINATE GEOMETRY

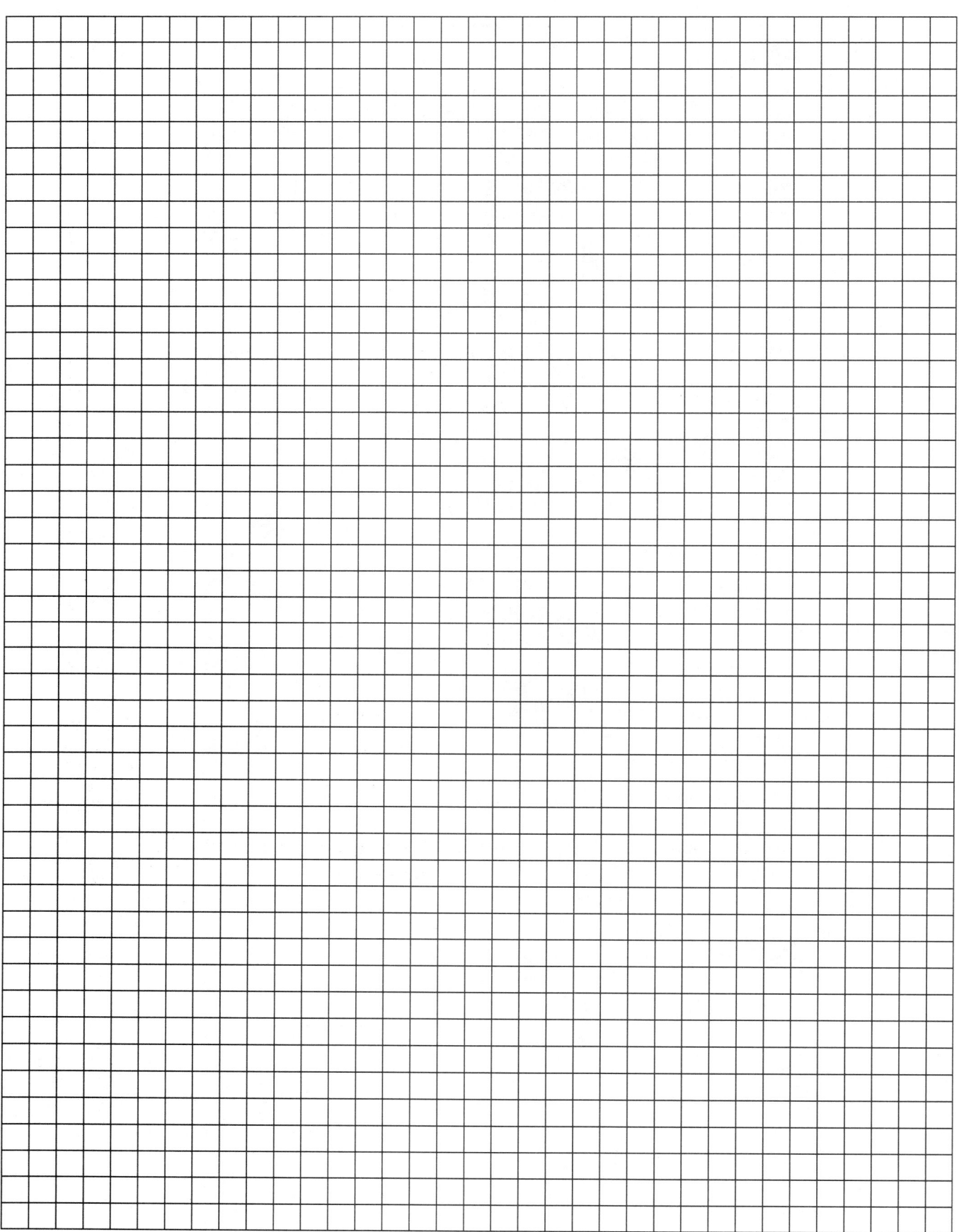

Chapter 6
TRANSFORMATIONAL GEOMETRY

Translations

> **REMEMBER**
>
> A **translation** is a type of **isometry**. This means that the image, after the translation, is congruent to the original object. A translation is a transformation that slides points the same distance in the same direction.
>
> The notation for a **translation** can be expressed as $T_{(-1,2)}$, $(x,y) \rightarrow (x-1, y+2)$ or as a vector $\langle -1,2 \rangle$. Each of these notations mean that a point is translated 1 unit to the left and 2 units up.
>
> A **translation** is a **direct isometry** because area, distance, angle measures, parallelism and orientation are preserved.
>
> Example: Describe the translation that maps the point $A(-3,2)$ to the point $A'(1,-1)$.
>
>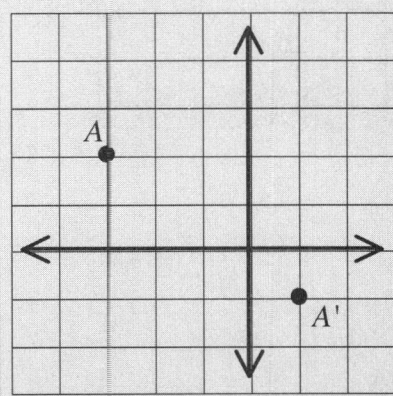
>
> Answer: The point A moves 4 units to the right and 3 units down. This can be described as $T_{(4,-3)}$, $(x,y) \rightarrow (x+4, y-3)$ or $\langle 4,-3 \rangle$.

1. Determine the coordinates of the image of the point $(5,-3)$ under $T_{(-2,-1)}$.	4. A translation maps the point $(-2,-5)$ to the point $(-4,-4)$. What is the image of $(1,4)$ under the same translation?
2. Determine the coordinates of the image of the point $(-8,-3)$ under the translation $(x,y) \rightarrow (x+4, y-1)$.	5. The image of a point under the translation $T_{(3,-4)}$ is $(-1,5)$. What are the coordinates of the point?
3. Determine the translation that maps the point $(-5,5)$ to the point $(7,1)$.	6. The image of a point under a reflection in the x-axis followed by the translation $T_{(-1,3)}$ is $(5,4)$. What are the coordinates of the point?

99

Chapter 6
TRANSFORMATIONAL GEOMETRY
Translations, Continued

7. Under which translation does all points of a figure remain invariant?

   (1) $T_{(1,1)}$ (3) $T_{(1,-1)}$
   (2) $T_{(-1,-1)}$ (4) $T_{(0,0)}$

   _____

8. Which of the following graphs demonstrates a translation?

   (1) 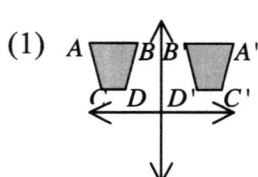 (3)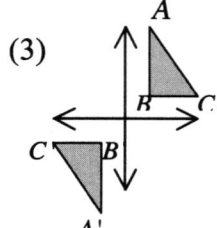

   (2) 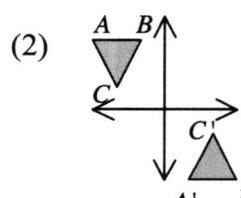 (4)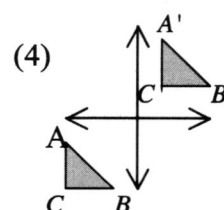

   _____

9. Which set of points maps $\overline{AB}$ to $\overline{A'B'}$ under a translation?

   (1) $A(2,3) \to A'(4,9)$
   $B(4,-1) \to B'(16,1)$

   (2) $A(2,3) \to A'(-2,3)$
   $B(4,-1) \to B'(-4,-1)$

   (3) $A(2,3) \to A'(3,2)$
   $B(4,-1) \to B'(-1,4)$

   (4) $A(2,3) \to A'(4,1)$
   $B(4,-1) \to B'(6,-3)$

   _____

10. Determine the translation that maps $\triangle ABC$ to $\triangle A'B'C'$.

    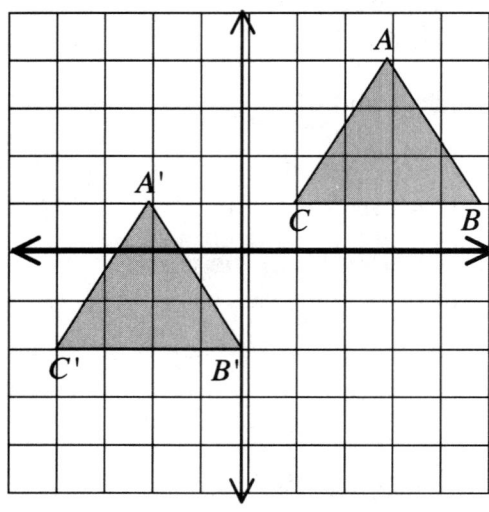

    (1) $T_{(-3,-3)}$ (3) $T_{(-3,3)}$
    (2) $T_{(5,3)}$ (4) $T_{(-5,-3)}$

    _____

11. Which of the following is the image of the point $(-5,7)$ after a translation of $T_{(-3,2)}$ followed by a translation of $T_{(2,3)}$ followed by a rotation of $R_{(O,90°)}$?

    (1) $(-12,-6)$ (3) $(-7,6)$
    (2) $(-12,0)$ (4) $(12,6)$

    _____

Chapter 6
TRANSFORMATIONAL GEOMETRY

Translations, (continued)

12. On the graph paper to the right, draw and label $\triangle LAG$ so that its' vertices are the points $L(-5,6)$, $A(-1,4)$ and $G(-3,1)$.

    Draw and label $\triangle L'A'G'$ after $T_{(7,-7)}$ of $\triangle LAG$.

    Is $\triangle L'A'G'$ also the image of $\triangle LAG$ after $R_{(O,180°)}$. Justify your answer.

13. On the graph paper to the right, draw and label $\square URSO$ so that its' vertices are the points $U(1,4)$, $R(3,4)$, $S(3,-1)$ and $O(1,-1)$.

    Draw and label $\square U'R'S'O'$ after a translation of $\square URSO$ that maps $(x,y) \rightarrow (x-4,y)$.

    Use the concept of orientation to justify that $\square U'R'S'O'$ is not the image of $\square URSO$ after a reflection in the y-axis.

14. On the graph paper to the right, draw and label $\triangle TAK$ so that its' vertices are the points $T(2,1)$, $A(4,5)$ and $K(6,1)$.

    Draw and label $\triangle T'A'K'$ after $T_{(x,y-6)}$ of $\triangle TAK$.

    Draw and label $\triangle T''A''K''$ after reflecting $\triangle T'A'K'$ in the y-axis.

    Is there a single translation that maps $\triangle TAK$ to $\triangle T''A''K''$? If so, determine the translation. If not, give a reason for why this translation is impossible.

101

Chapter 6
TRANSFORMATIONAL GEOMETRY
Line Reflections

> **REMEMBER**
>
> A **line reflection** is a type of **isometry**. This means that the image, after the line reflection, is congruent to the original object.
>
> A line reflection is the image that is mirrored over a line.
>
> The **line of reflection** is the **perpendicular bisector** of the line segment connecting a point to its' image point. Points on the line of reflection are invariant under the reflection.
>
> Under a line reflection area, distance angle measures, and parallelism are preserved, but orientation switches. This means that a line reflection is a type of **opposite isometry** because if the object was oriented in a clockwise manner, the image is oriented in a counterclockwise manner.
>
> Below are some standard line reflection formulas about the different lines in the plane:
>
> | Line of Reflection | Notation | Formula | Example | Answer |
> |---|---|---|---|---|
> | x-axis or $y=0$ | $r_{y=0}(x,y)$ | $(x,y) \to (x,-y)$ | $(3,5)$ | $(3,-5)$ |
> | y-axis or $x=0$ | $r_{x=0}(x,y)$ | $(x,y) \to (-x,y)$ | $(2,1)$ | $(-2,1)$ |
> | $y=x$ | $r_{y=x}(x,y)$ | $(x,y) \to (y,x)$ | $(-3,4)$ | $(4,-3)$ |
> | $y=-x$ | $r_{y=-x}(x,y)$ | $(x,y) \to (-y,-x)$ | $(4,-2)$ | $(2,-4)$ |
>
> **Example:** In the figure below, what is the image of $\overline{AB}$ after a reflection in line ℓ?
>
>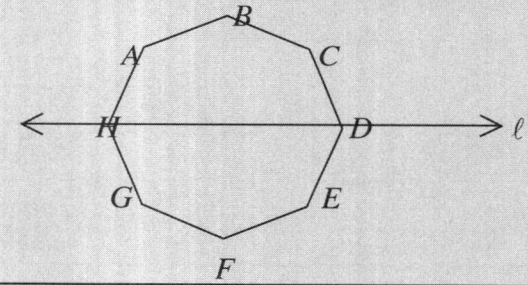
>
> Answer: $\overline{GF}$

1. What is the image of $(4,-7)$ after a reflection in the x-axis?

   (1) $(4,7)$ (3) $(-4,7)$
   (2) $(-4,-7)$ (4) $(7,4)$

   _____

2. What is the image of $(-6,-1)$ after a reflection in the line $y=x$?

   (1) $(6,1)$ (3) $(-1,-6)$
   (2) $(1,6)$ (4) $(-6,1)$

   _____

2. What is the result of $r_{y-axis}(-3,10)$?

   (1) $(-3,-10)$ (3) $(3,10)$
   (2) $(3,-10)$ (4) $(10,-3)$

   _____

4. What is the result of $r_{y=-x}(5,-1)$?

   (1) $(-1,5)$ (3) $(-5,1)$
   (2) $(5,1)$ (4) $(1,-5)$

   _____

Chapter 6
TRANSFORMATIONAL GEOMETRY
Line Reflections, (continued)

5. In the picture below, circle two figures which are reflections of each other and label the line of reflection.

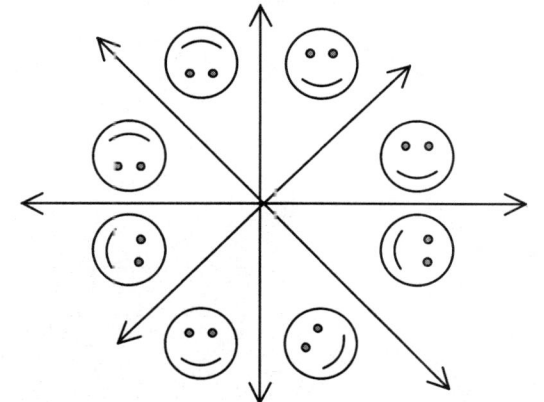

6. State a reason for why $\overline{A'B'}$ is **not** the reflection of $\overline{AB}$ in line ℓ.

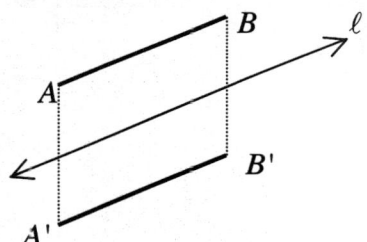

7. Which segment is the image of $\overline{IJ}$ after a reflection over the x-axis and then the y-axis?

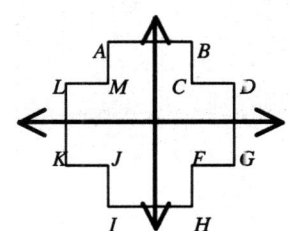

(1) $\overline{AM}$ (3) $\overline{HF}$
(2) $\overline{BC}$ (4) $\overline{CD}$

8. The reflection of $\triangle ABC$ in line ℓ is $\triangle A'B'C'$. Determine the area of $\triangle ABC$.

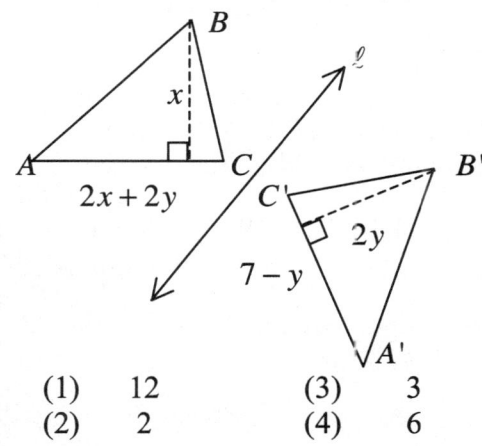

(1) 12   (3) 3
(2) 2   (4) 6

9. The graph of which equation is the image of the graph of the line $y = -\frac{1}{2}x + 3$ after a reflection in the y-axis?

(1) $y = \frac{1}{2}x - 3$ (3) $y = \frac{1}{2}x + 3$

(2) $y = -2x - 3$ (4) $y = -\frac{1}{2}x - 3$

10. On the axes below sketch a pentagon whose image after a reflection over the x-axis is the same pentagon.

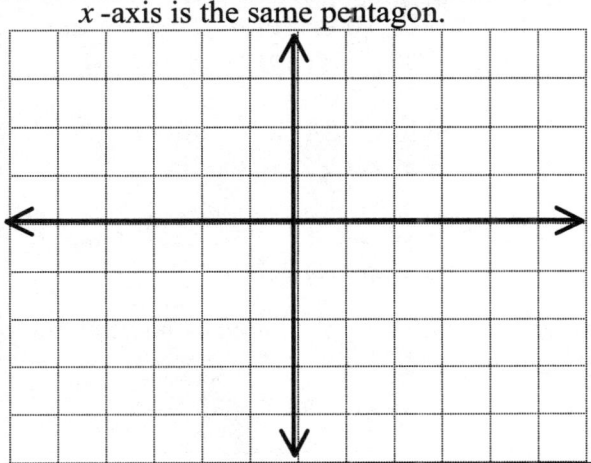

Chapter 6

TRANSFORMATIONAL GEOMETRY

Line Reflections, (continued)

11. On the graph paper to the right, draw and label $\triangle TOM$ so that its' vertices are the points $T(2,1)$, $O(5,1)$ and $M(5,4)$.

    Draw and label $\triangle T'O'M'$ after a reflection of $\triangle TOM$ in the y-axis.

    Draw and label $\triangle T''O''M''$ after a reflection of $\triangle T'O'M'$ in the x-axis.

    Name a single transformation that would map $\triangle TOM$ onto $\triangle T''O''M''$.

12. On the graph paper to the right, draw and label parallelogram $MIKE$ so that its' vertices are the points $M(2,0)$, $I(5,3)$, $K(4,6)$ and $E(1,3)$.

    Draw and label $\square M'I'K'E'$ after a reflection of $\square MIKE$ in the x-axis.

    Draw and label $\square M''I''K''E''$ after $R_{(O,180°)}$ of $\square M'I'K'E'$.

    Name a single transformation that would map $\square MIKE$ onto $\square M''I''K''E''$.

13. On the graph paper to the right, draw and label $\triangle AMY$ so that its vertices are the points $A(2,1)$, $M(5,0)$ and $Y(0,5)$.

    Draw and label $\triangle A'M'Y'$ after a reflection of $\triangle AMY$ in the line $y=-x$.

    What type of quadrilateral is $MYY'M'$? Justify your answer.

    Is $\triangle MAM'$ a right triangle? Justify your answer.

Chapter 6

TRANSFORMATIONAL GEOMETRY

Line Reflections, (continued)

14. Which of the following points is the reflection of the point $(-5,2)$ over the line $y=4$?

    (1) $(5,2)$ (3) $(-5,-2)$
    (2) $(-5,6)$ (4) $(13,2)$

15. Which of the following points is the reflection of the point $(-2,-3)$ over the line $y=1$?

    (1) $(-2,1)$ (3) $(-2,5)$
    (2) $(4,-3)$ (4) $(0,-3)$

16. Which of the following points is the reflection of the point $(3,4)$ over the line $y=-1$?

    (1) $(3,-6)$ (3) $(-5,4)$
    (2) $(3,-2)$ (4) $(-1,4)$

17. Which of the following points is the reflection of the point (x,y) over the line $y=k$?

    (1) $(x,2k+y)$ (3) $(x,2k-y)$
    (2) $(x,2y+k)$ (4) $(x,2y-k)$

18. Which of the following equations is the line of reflection for $\square JACK$ and $\square J'A'C'K'$ (shown to the right)?

    (1) $x=1$ (3) $y=1$
    (2) $x=2$ (4) $y=2$

19. Which of the following points is the reflection of the point $(5,1)$ over the line $x=2$?

    (1) $(5,3)$ (3) $(-1,1)$
    (2) $(1,1)$ (4) $(-5,1)$

20. Which of the following points is the reflection of the point $(2,-6)$ over the line $x=-1$?

    (1) $(-4,-6)$ (3) $(0,-6)$
    (2) $(2,4)$ (4) $(-4,6)$

21. Which of the following points is the reflection of the point $(-4,-2)$ over the line $x=1$?

    (1) $(-4,4)$ (3) $(-4,0)$
    (2) $(6,-2)$ (4) $(2,-2)$

22. Which of the following points is the reflection of the point (x,y) over the line $x=h$?

    (1) $(2h+x,y)$ (3) $(2h-x,y)$
    (2) $(2x+h,y)$ (4) $(2x-h,y)$

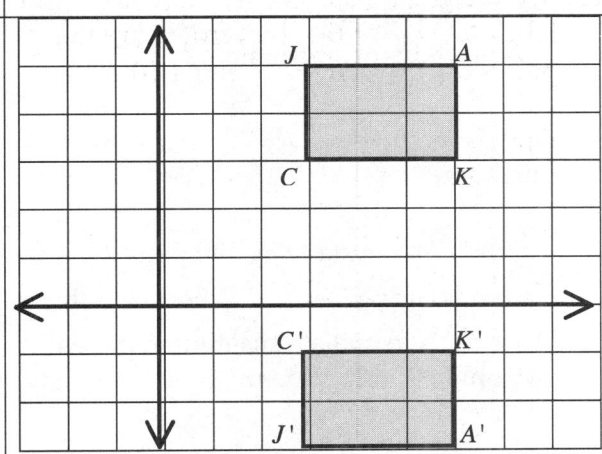

Chapter 6

TRANSFORMATIONAL GEOMETRY

Glide Reflections

> **REMEMBER**
>
> A **glide reflection** is a type of **isometry**. A **glide reflection** combines a **line reflection** with a **translation** in the <u>same direction as the line</u>. With a glide reflection, it does not matter which transformation is done first; the resultant image will be the same.
>
> The following properties are **invariant** (are preserved) under a **glide reflection**: **distance, area, angle measures, colinearity and parallelism.** Since the **orientation** (clockwise or counterclockwise) is switched under a glide reflection, a glide reflection is known as an **opposite isometry**.
>
>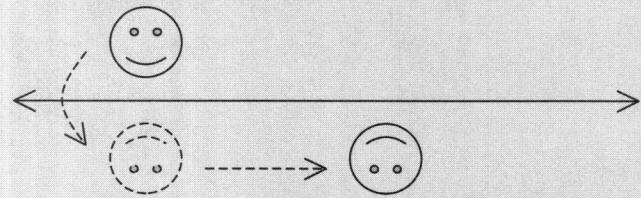
>
> Example: What are the coordinates of the point $(3,-2)$ after a reflection in the x-axis followed by $T_{(4,0)}$?
>
> Answer: First, reflect the point in the x-axis to get the point $(3,2)$. Then translate the point following the rule $(x,y) \rightarrow (x+4,y)$ to get the point $(7,2)$.

1. What is the image of the point $(-5,2)$ after a reflection in the y-axis followed by the translation $(x,y) \rightarrow (x,y+3)$?	4. A glide reflection maps the point $(8,1)$ to the point $(-5,-4)$ using a reflection over the line $y=-x$. What translation completes this glide reflection?
2. What is the image of the point $(2,-6)$ after a reflection in the line $y=x$ followed by the translation $T_{(-1,-1)}$?	5. A glide reflection maps the point $(-5,-3)$ to the point $(1,1)$. Determine one of the glide reflections that will accomplish this transformation.
3. A glide reflection maps the point $(2,4)$ to the point $(-2,-4)$. Determine a line of reflection and a translation that could accomplish this glide reflection.	6. The slope of the line of reflection and the ratio of the vertical translation to the horizontal translation are equal. Why must this be true for a glide reflection?

Chapter 6

TRANSFORMATIONAL GEOMETRY
Glide Reflections, (continued)

7. In the accompanying figure, $\triangle W'E'N'$ is the image of $\triangle WEN$ under a glide reflection. Draw the line of reflection and indicate the direction of the translation that completes this transformation.

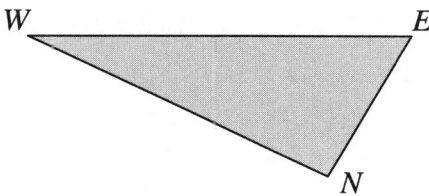
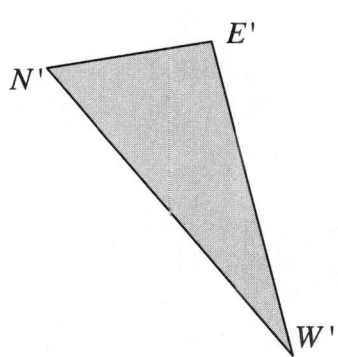

8. In the accompanying diagram, $\square S'E'A'N'$ is the image of $\square SEAN$ under a glide reflection. Draw the line of reflection and indicate the direction of the translation that complete this transformation.

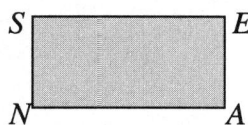
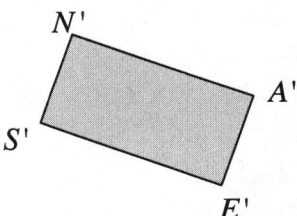

9. In the accompany diagram, $\square T'O'R'Y'$ is the image of $\square TORY$ under a glide reflection. Draw the line of reflection and indicate the direction of the translation that completes this transformation.

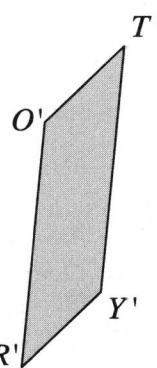

Chapter 6

TRANSFORMATIONAL GEOMETRY
Glide Reflections, (continued)

10. On the graph paper to the right, draw and label $\triangle MRA$ with vertices $M(-5,1)$, $R(-3,5)$ and $A(-2,2)$.

    Draw and label $\triangle M'R'A'$ after a reflection in the y-axis.

    Draw and label $\triangle M''R''A''$ after a translation of $T_{(0,-5)}$.

    Complete the chart to the right for the coordinates of $\triangle MRA$ and $\triangle M''R''A''$.

    Complete the translation below that maps $\triangle MRA$ directly on to $\triangle M''R''A''$.

    $(x,y) \to (\ -x\ ,\ y-5\ )$

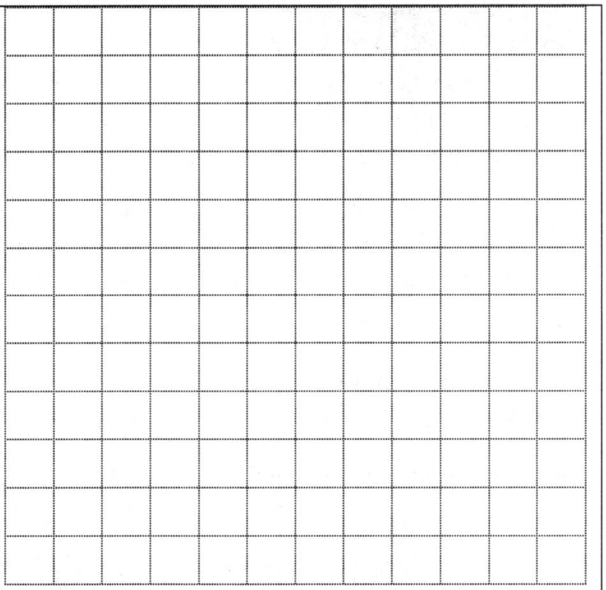

$\triangle MRA$	M	R	A
Point	(−5, 1)	(−3, 5)	(−2, 2)
$\triangle M''R''A''$	M''	R''	A''
Point	(5, −4)	(3, 0)	(2, −3)

11. On the graph paper to the right, draw and label $\triangle HAN$ with vertices $H(2,2)$, $A(4,-2)$ and $N(1,-3)$.

    Draw and label $\triangle H'A'N'$ after a reflection in the the line $y=-x$.

    Draw and label $\triangle H''A''N''$ after a translation of $T_{(-4,4)}$.

    Complete the chart to the right for the coordinates of $\triangle HAN$ and $\triangle H''A''N''$.

    Complete the translation below that maps $\triangle HAN$ directly on to $\triangle H''A''N''$.

    $(x,y) \to (\ -y-4\ ,\ -x+4\ )$

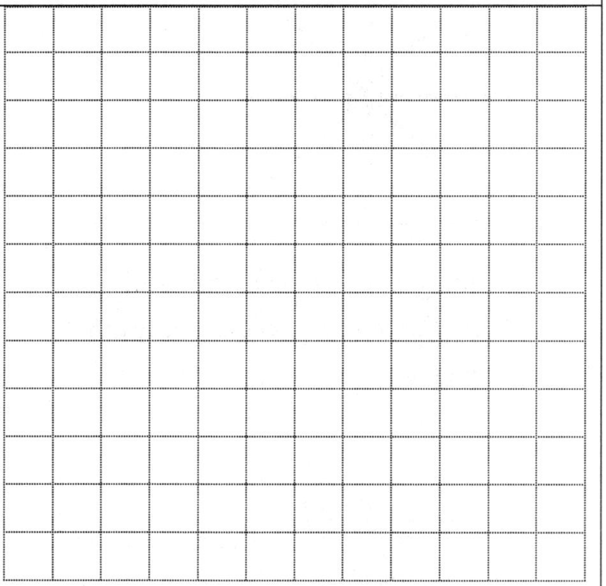

$\triangle HAN$	H	A	N
Point	(2, 2)	(4, −2)	(1, −3)
$\triangle H''A''N''$	H''	A''	N''
Point	(−6, 2)	(−2, 0)	(−1, 3)

Chapter 6

TRANSFORMATIONAL GEOMETRY

Rotations

> **REMEMBER**
>
> A **rotation** is a type of isometry. This means that the image, after the rotation, is congruent to the original object. The center of the rotation is an invariant point because this is a point that does not change location (or it maps onto itself).
>
> Rotations are expressed by the number of degrees of rotation and can be either clockwise or counterclockwise. A counterclockwise rotation is considered a positive rotation and a clockwise rotation is considered a negative rotation.
>
> A rotation is often written in the form $R_{(Center, Degrees)}$, where the **center** is the point of the 'wheel' around which the point will rotate and degrees is the number of degrees of rotation. A positive number of degrees represents a counterclockwise rotation.
>
> Below are some standard point rotation formulas about the origin:
>
> | Notation | Formula | Example | Answer |
> |---|---|---|---|
> | $R_{(O, 90°)}$ | $(x, y) \rightarrow (-y, x)$ | $(3, 5)$ | $(-5, 3)$ |
> | $R_{(O, 180°)}$ | $(x, y) \rightarrow (-x, -y)$ | $(2, 1)$ | $(-2, -1)$ |
> | $R_{(O, -90°)}$ | $(x, y) \rightarrow (y, -x)$ | $(-3, 4)$ | $(4, 3)$ |
>
> Example: What is the image if $\overline{AO}$ is rotated 135° counterclockwise about point O?
>
> Answer: $\overline{GO}$
>
>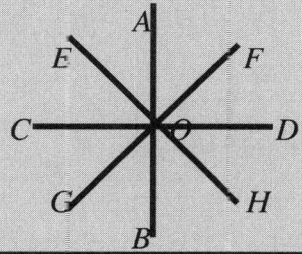

1. Which of the following coordinates is the image of $(-1, -7)$ after a 90° clockwise rotation about the origin?

   (1) $(-7, -1)$ (3) $(-7, 1)$
   (2) $(7, -1)$ (4) $(7, 1)$

2. Which of the following coordinates is the image of $(2, -5)$ after $R_{(O, 90°)}$?

   (1) $(-5, 2)$ (3) $(5, -2)$
   (2) $(-5, -2)$ (4) $(5, 2)$

3. Which of the following coordinates is the image of $(8, 4)$ after a rotation of 180° about the origin?

   (1) $(4, 8)$ (3) $(-4, -8)$
   (2) $(-8, -4)$ (4) $(-8, 4)$

4. Which of the following coordinates is the image of $(-7, 3)$ after $R_{(O, -90°)}$?

   (1) $(-7, -3)$ (3) $(-3, -7)$
   (2) $(3, 7)$ (4) $(3, -7)$

Chapter 6

TRANSFORMATIONAL GEOMETRY

Rotations, (continued)

5. Estimate the angle of rotation:

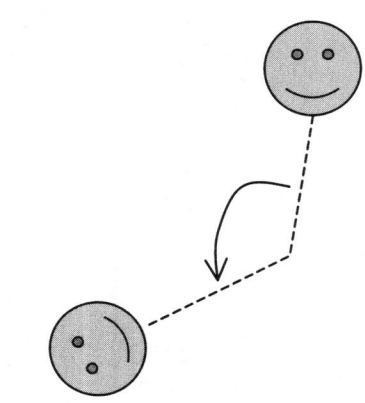

(1) 45° (3) 135°
(2) 90° (4) 180°

_____

6. What angle of rotation about O would map $\triangle ABO$ to $\triangle DEO$?

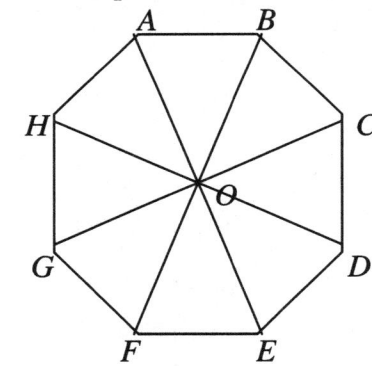

(1) 135° (3) 225°
(2) −225° (4) 45°

_____

7. The graph of which equation is the image of the graph of the line $y = 2x - 4$ after $R_{(O, 90°)}$?

(1) $y = 2x + 4$ (3) $y = -2x + 2$
(3) $y = -\frac{1}{2}x + 2$ (4) $y = \frac{1}{2}x - 4$

_____

8. Which of the following represents the image of $\triangle ABC$ after $R_{(90°)}$?

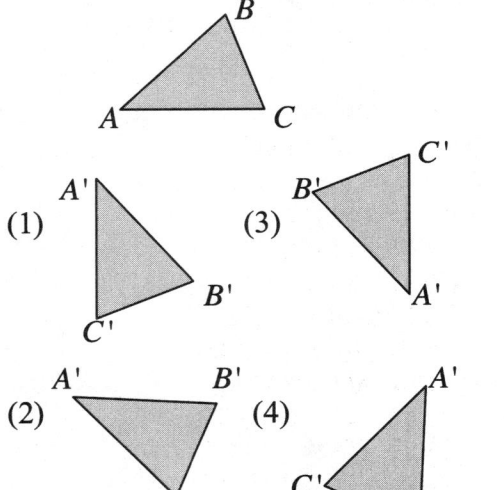

_____

9. Which of the following represents the location and orientation of the letter M after $R_{(O, 90°)}$?

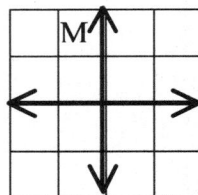

(1) 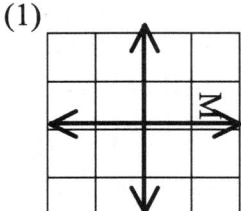   (3)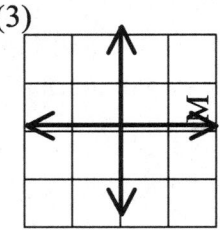

(2) 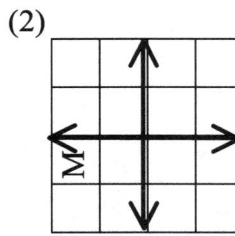   (4)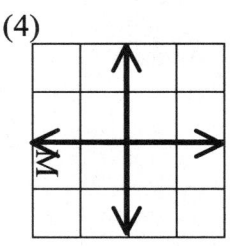

_____

Chapter 6

TRANSFORMATIONAL GEOMETRY

Rotations, (continued)

10. On the graph paper to the right, draw and label $\triangle DAN$ so that its' vertices are the points $D(5,0)$, $A(1,4)$ and $N(-2,1)$.

    Draw and label $\triangle D'A'N'$ after a $R_{(O,90°)}$ of $\triangle DAN$.

    Draw and label $\triangle D''A''N''$ after transforming $\triangle D'A'N'$ under $R_{(O,180°)}$.

    Symbolically, write the transformation that would map $\triangle DAN$ onto $\triangle D''A''N''$ after one transformation.

11. On the graph paper to the right, draw and label quadrilateral $LISA$ so that its' vertices are the points $L(2,1)$, $I(6,3)$, $S(4,5)$ and $A(1,2)$.

    Draw and label quadrilateral $L'I'S'A'$ after a $R_{(0,-90°)}$ of quadrilateral $LISA$.

    Draw and label quadrilateral $L''I''S''A$ after a $R_{(A,90°)}$ of quadrilateral $LISA$.

12. On the graph paper to the right, draw and label $\triangle RMH$ so that its vertices are the points $R(0,1)$, $M(3,3)$ and $H(-3,3)$.

    Draw and label $\triangle R'M'H$ after a $R_{(H,-90°)}$ of $\triangle RMH$.

    Draw and label $\triangle R''H''M'$ after a $R_{(M',-90°)}$ of $\triangle R'M'H$.

    Draw and label $\triangle R'''MH''$ after a $R_{(H'',-90°)}$ of $\triangle R''H''M'$.

    What type of polygon is figure $MHR\ M'H''R'''$?

Chapter 6

TRANSFORMATIONAL GEOMETRY

Dilations

> **REMEMBER**
>
> A **dilation** is a type of **transformation** that results in an image that is **similar** to the original figure. A **dilation** is **not** an **isometry** because the image is not congruent to the original figure (unless the factor of dilation is 1 or −1).
>
> A **dilation** can either **enlarge** or **reduce** an object. The amount a figure is enlarged or reduced is called the **scale factor** or the **factor of dilation**.
>
> If the scale factor is greater than 1, the image will be larger than the object.
> If the scale factor is less than 1, the image will be smaller than the object.
>
> Angle measurement, orientation, parallelism and collinearity are **invariant** (stay the same) under a dilation. Area and distance are **not invariant** and will change under a dilation.
>
> The notation for a dilation is D_k, where k is the value of the scale factor. To dilate a point, use the formula $D_k(x,y) = (kx, ky)$.
>
> Although the **origin** is often the **center of a dilation** on a coordinate axis, to find the center of a dilation, draw lines connecting points on the original figure to their respective image points on the dilated figure. The point at which these lines are concurrent (meet) is the center of the dilation.
>
> **Example:** What is the image of the point $(-9, 3)$ under the dilation $D_{\frac{4}{3}}$?
>
> **Answer:** $D_{\frac{4}{3}}(-9, 3) = \left(\frac{4}{3} \cdot -9, \frac{4}{3} \cdot 3\right) = (-12, 4)$

1. Which of the following points is the image of $(2, -6)$ under the dilation D_2 centered at the origin?

   (1) $(1, -3)$ (3) $(4, -4)$
   (2) $(2, -6)$ (4) $(4, -12)$

2. What is the image of $(8, -10)$ under the dilation $D_{\frac{3}{2}}$ centered at the origin?

3. Which dilation centered at the origin maps $(-3, 7)$ to $(6, -14)$?

   (1) D_2 (3) $D_{-\frac{1}{2}}$
   (2) D_{-2} (4) D_9

4. The perimeter of $\triangle ABC$ is 108 inches and the perimeter of $\triangle A'B'C'$, its image under a dilation centered at the origin is 36 inches. This dilation can be written:

   (1) $D_{\frac{1}{3}}$ (3) $D_{-\frac{1}{3}}$
   (2) D_3 (4) D_{-3}

5. If the image of $(3, -5)$ is $(x, 15)$, what is the value of x?

6. If the image under a dilation of D_2 centered at the origin is $(4, -10)$, which of the following is the original point?

   (1) $(6, -8)$ (3) $(2, -12)$
   (2) $(2, -5)$ (4) $(8, -20)$

Chapter 6

TRANSFORMATIONAL GEOMETRY

Dilations, (continued)

7. On the graph paper to the right, draw and label $\triangle MAD$ with vertices $M(-1,1)$, $A(-2,-1)$ and $D(-3,-3)$.

   Draw and label $\triangle M'A'D'$, the image of $\triangle MAD$ after a dilation centered at the origin of D_2.

   Draw and label $\triangle M''A''D''$, the image of $\triangle M'A'D'$ after a dilation centered at the origin of $D_{-\frac{3}{2}}$.

   Name the single transformation that will map $\triangle MAD$ onto $\triangle M''A''D''$.

8. On the graph paper to the right, draw and label $\triangle JEN$ with vertices $J(0,0)$, $E(1,3)$ and $N(3,1)$.

   Draw and label $\triangle J'E'N'$, the image of $\triangle JEN$ after a dilation centered at the origin of D_2.

   Draw and label $\triangle J''E''N''$, the image of $\triangle J'E'N'$ after a dilation centered at the origin of D_{-1}.

   Name a single transformation that will map $\triangle JEN$ onto $\triangle J''E''N''$.

9. On the graph paper to the right, draw and label $\square NICH$ with vertices $N(2,-1)$, $I(4,-2)$, $C(6,-5)$, and $H(4,-4)$.

   Draw and label $\square N'I'C'H'$, the image of $\square NICH$ after a dilation centered at the origin of D_{-1}.

   Draw and label $\square N''I''C''H''$, the image of $\square N'I'C'H'$ after a reflection in the y-axis.

   Name a single transformation that will map $\square NICH$ onto $\square N''I''C''H''$.

Chapter 6

TRANSFORMATIONAL GEOMETRY

Dilations, (continued)

10. Right $\triangle R'E'A'$ is the image of right $\triangle REA$ after a dilation of D_2 with center P. E and E' are right angles. The measure of $\overline{RE} = x$, $\overline{R'E'} = y - 2$, $\overline{AE} = 4$ and $\overline{A'E'} = 2x + 2$.

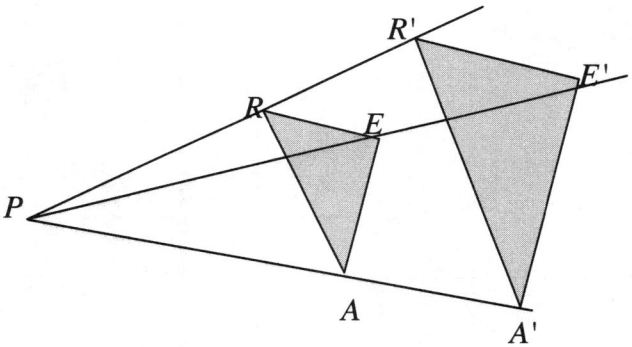

Determine the values of x and y and the length of $\overline{A'B'}$.

11. Isosceles trapezoid $J'E'S'D'$ is the image of isosceles trapezoid $JESD$ after a dilation of $D_{\frac{1}{2}}$ with center P. The measure of $\overline{JE} = 4$, $\overline{ES} = x + 1$, $\overline{AJ} = 5x + 7$, $\overline{E'S'} = y - 1$ and $\overline{A'J'} = 4y$.

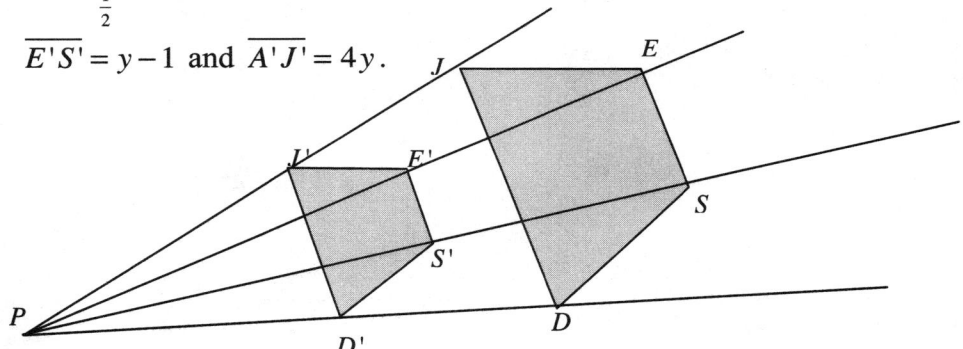

Determine the perimeter of isosceles trapezoid $J'E'S'D'$.

Chapter 6

TRANSFORMATIONAL GEOMETRY

Dilations, (continued)

12. In the diagram below, $\triangle A'B'C'$ is the image of $\triangle ABC$ under a dilation with center P. Determine the location of point P. Show your method and justify your answer using the property(ies) of dilations.

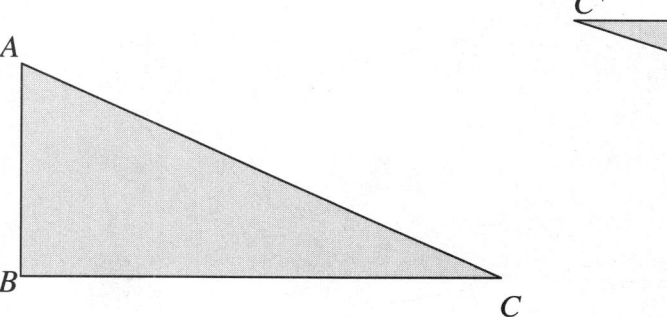
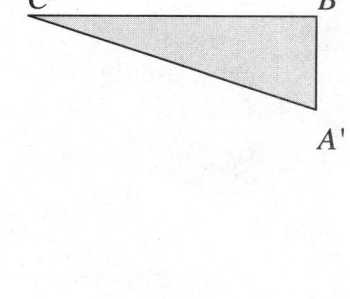

13. In the diagram to the below, $\triangle A'B'C'$ is the image of $\triangle ABC$ under a dilation with center P.

    Determine the coordinates of point P.

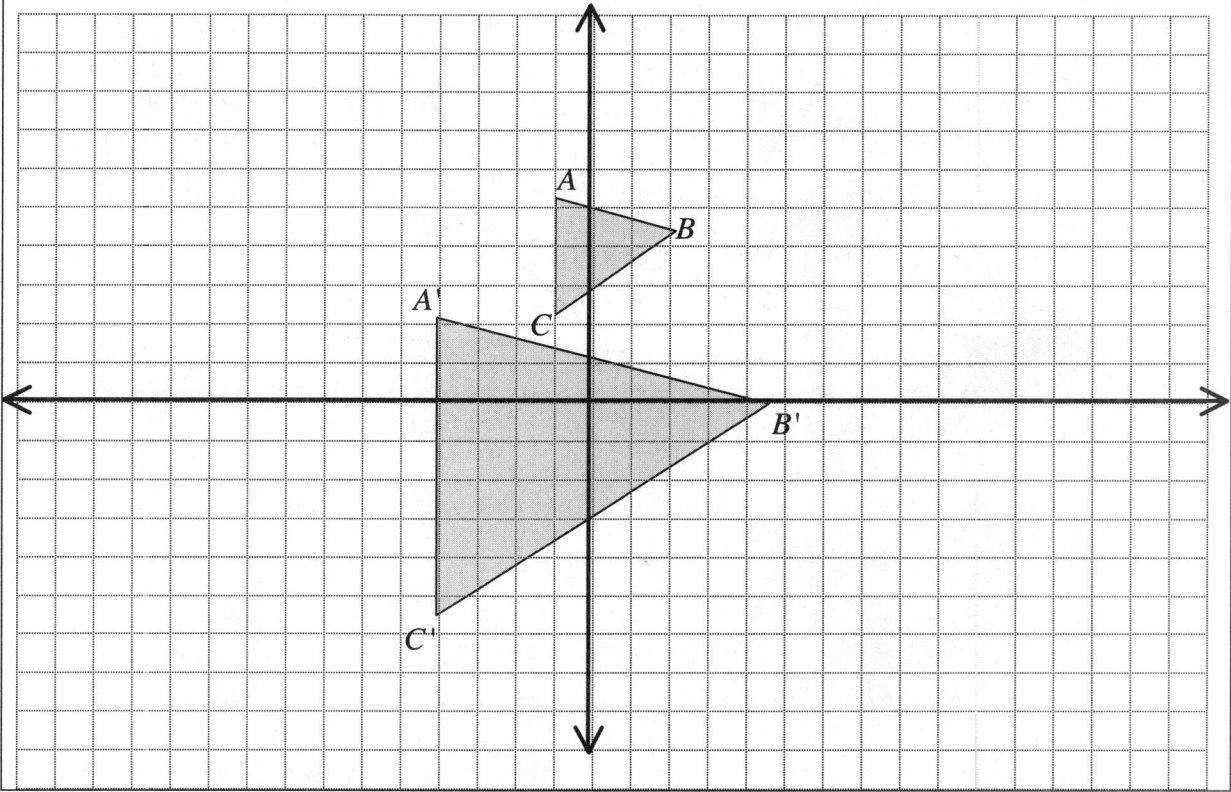

CHAPTER 6

TRANSFORMATIONAL GEOMETRY

Compositions of Isometries and Dilations

> **REMEMBER**
>
> A **composition of isometries and dilations** involves two or more transformations. The first transformation maps to an image and this image undergoes the second transformation. A glide reflection is an example of a composition of transformations.
>
> The notation for a composition of transformations is similar to the notation for the composition of functions. Similar to the composition of functions, the transformation that is written second is the first to be performed.
>
> **Example:** What is the image of the point $A(-5,8)$ after the transformation $R_{(O,90°)} \circ r_{y-axis} \circ T_{(4,-6)}$?
>
> **Answer:** (1) Translate point A using the rule $(x,y) \to (x+4, y-6)$ to obtain $A'(-1,2)$.
> (2) Reflect point A' using the rule $(x,y) \to (-x,y)$ to obtain $A''(1,2)$.
> (3) Rotate point A'' using the rule $(x,y) \to (-y,x)$ to obtain $A'''(-2,1)$.
> The answer is the point $(-2,1)$.

1. What is the image of the point $(9,5)$ after $T_{(-2,3)} \circ r_{x-axis}$?

2. What is the result of the transformation $r_{x-axis} \circ R_{(O,180°)}$ on the point $(-1,7)$?

3. Which of the following is the image of the point $(-6,-2)$ after $D_2 \circ T_{(4,3)}$?

   (1) $(1,2)$ (3) $(-8,-1)$
   (2) $(-2,1)$ (4) $(-4,2)$

4. Which of the following is the image of the point $(5,-3)$ after $R_{(O,-90°)} \circ r_{y=x}$?

   (1) $(-3,5)$ (3) $(5,3)$
   (2) $(-5,3)$ (4) $(-5,-3)$

5. Two transformations are performed on the point $(3,-2)$ with a resulting image point of $(7,1)$. If the second was a reflection over the x-axis, determine a possible first transformation.

6. What are the two transformations that map the point $A(-6,-9)$ to the point $A'(-3,-12)$ to the point $A''(-1,-4)$.

7. Which of the following is the rule for the composition of transformations determined in question #6?

   (1) $(x,y) \to \left(\dfrac{1}{3}x+3, \dfrac{1}{3}y-3\right)$
   (2) $(x,y) \to \left(\dfrac{1}{3}x+1, \dfrac{1}{3}y-1\right)$
   (3) $(x,y) \to (3x+3, 3y-3)$
   (4) $(x,y) \to (3x+1, 3x-1)$

CHAPTER 6

TRANSFORMATIONAL GEOMETRY
Compositions of Isometries and Dilations, continued

8. Consider the figure below. Write, using proper notation, a composition of transformations that will map $\triangle ABC$ onto $\triangle A'B'C'$?

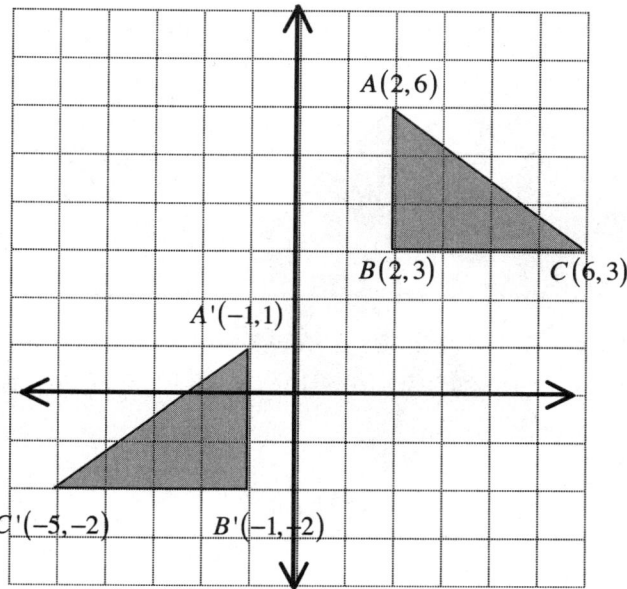

9. Consider the figure below. Write, using proper notation, a composition of transformations that will map $\triangle DEF$ onto $\triangle D'E'F'$?

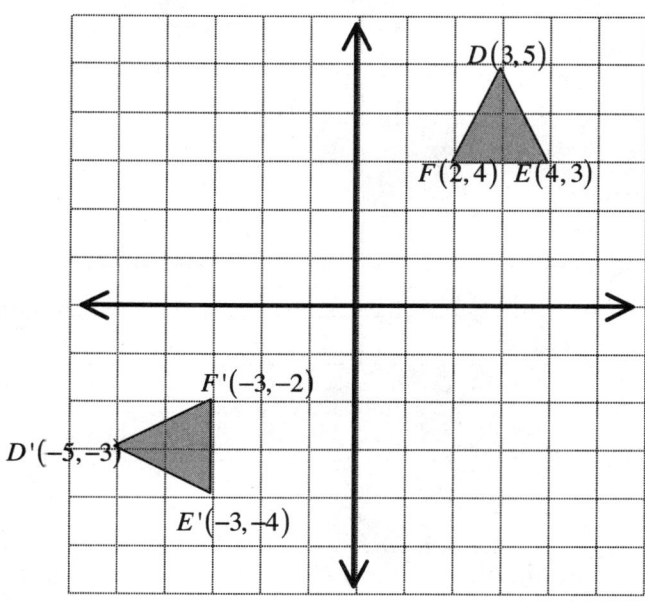

CHAPTER 6

TRANSFORMATIONAL GEOMETRY
Compositions of Isometries and Dilations, Continued

10. Consider the figure below. Write, using proper notation, a composition of transformations that will map $\square METS$ onto $\square M'E'T'S'$?

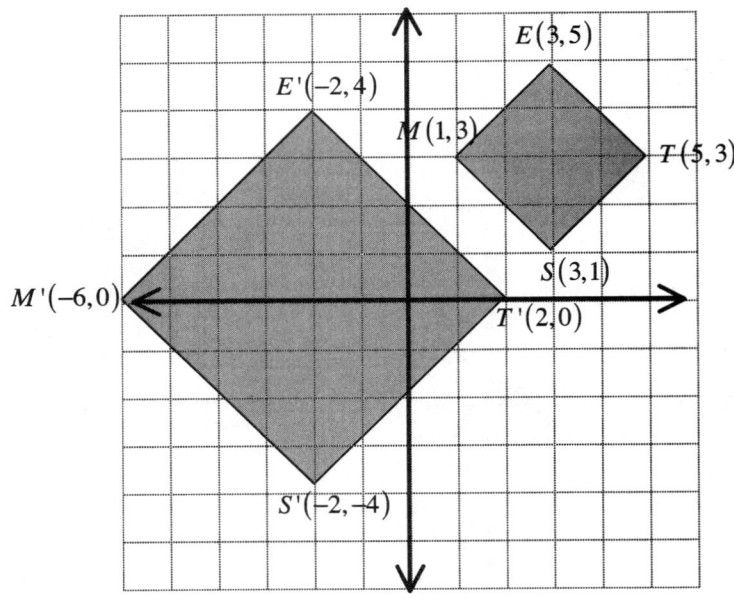

$T_{-8,-6} \circ D_2$

11. Consider the figure below. Write, using proper notation, a composition of transformations that will map $\square JETS$ onto $\square J'E'T'S'$?

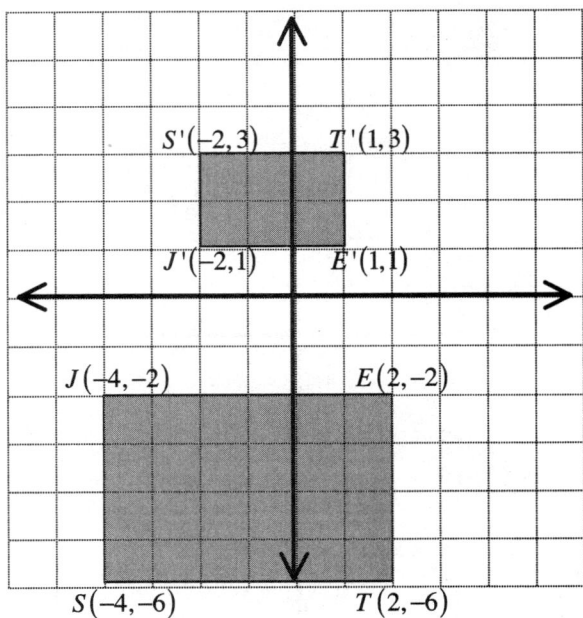

$D_{1/2} \circ r_{x\text{-axis}}$

118

CHAPTER 6

TRANSFORMATIONAL GEOMETRY
Compositions of Isometries and Dilations, Continued

12. Consider the figure below. Write, using proper notation, a composition of transformations that will map $\triangle LIN$ onto $\triangle L'I'N'$?

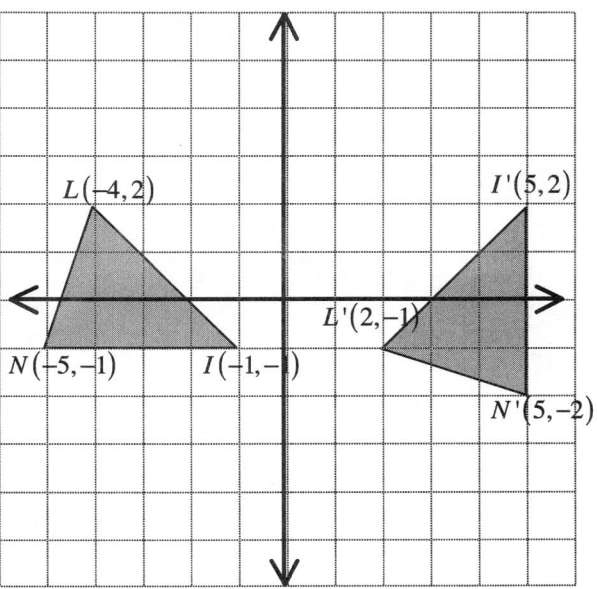

13. Consider the figure below. Write, using proper notation, a composition of transformations that will map $\triangle MOE$ onto $\triangle M'O'E'$?

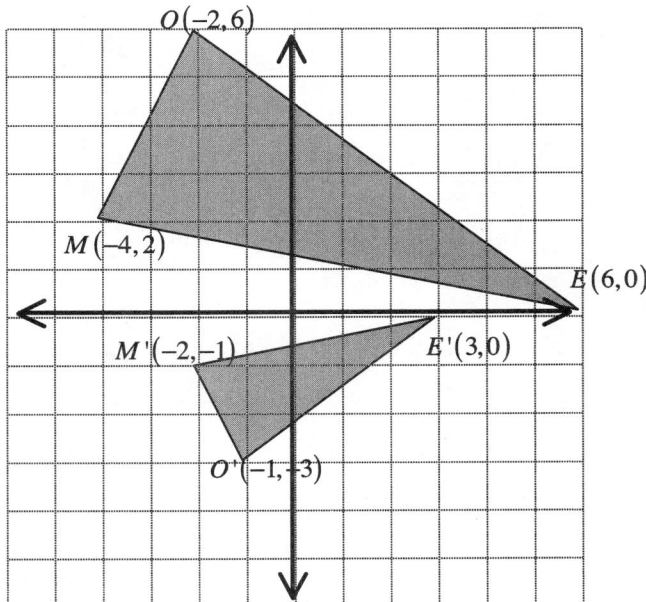

CHAPTER 6

TRANSFORMATIONAL GEOMETRY
Compositions of Isometries and Dilations, Continued

14. Consider trapezoid *KRIS* with coordinates as shown in the figure below.
    Draw and label $K"R"I"S"$, the image of *KRIS*, after the transformation $r_{x=4} \circ r_{x=-2}$.

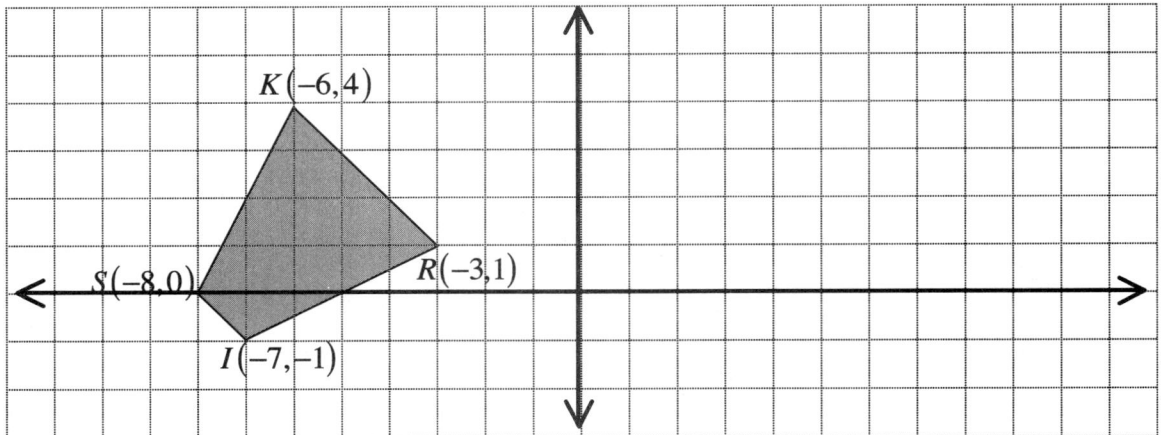

What single transformation is the reflection of *KRIS* over two parallel lines equivalent to? (Use proper transformation notation to express your answer.)

15. Consider $\triangle KAT$ with coordinates as shown in the figure below.
    Sketch $\triangle K"A"T"$, the image of $\triangle KAT$, after the transformation $r_m \circ r_\ell$.
    (The coordinate points of $\triangle K"A"T"$ will not be integral.)

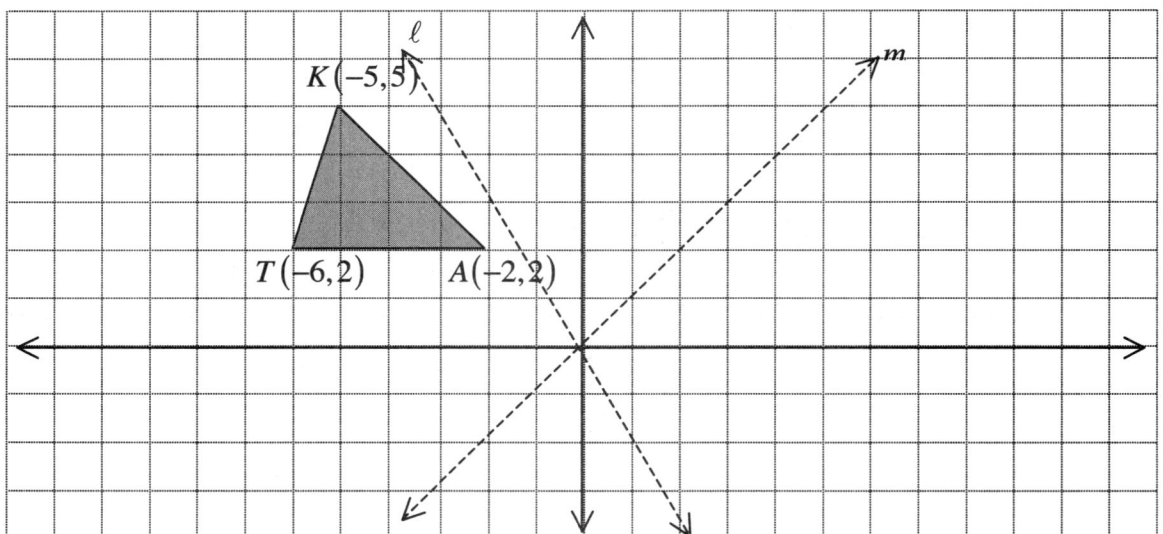

If the angle between lines ℓ and m is $60°$, use the properties of transformations to determine the single transformation that maps $\triangle KAT$ onto $\triangle K"A"T"$. (Use proper transformation notation to express your answer.)

CHAPTER 6

TRANSFORMATIONAL GEOMETRY
Compositions of Isometries and Dilations, Continued

16. Consider $\triangle ABC$ with coordinates as shown in the figure below.

    Draw and label $\triangle A"B"C"$, the image of $\triangle ABC$, after the transformation $R_{(O,-90°)} \circ r_{y=-x}$.

    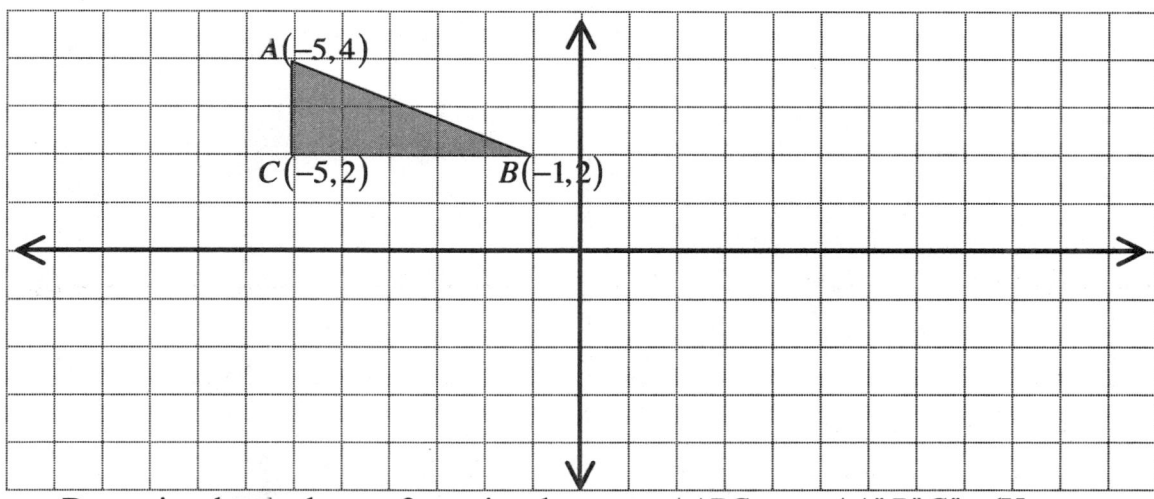

    Determine the single transformation that maps $\triangle ABC$ onto $\triangle A"B"C"$. (Use proper transformation notation to express your answer.)

17. Consider $\triangle ABC$ with coordinates as shown in the figure below.

    Draw and label $\triangle A"B"C"$, the image of $\triangle ABC$, after the transformation $R_{(O,-90°)} \circ r_{y=-x+1}$.

    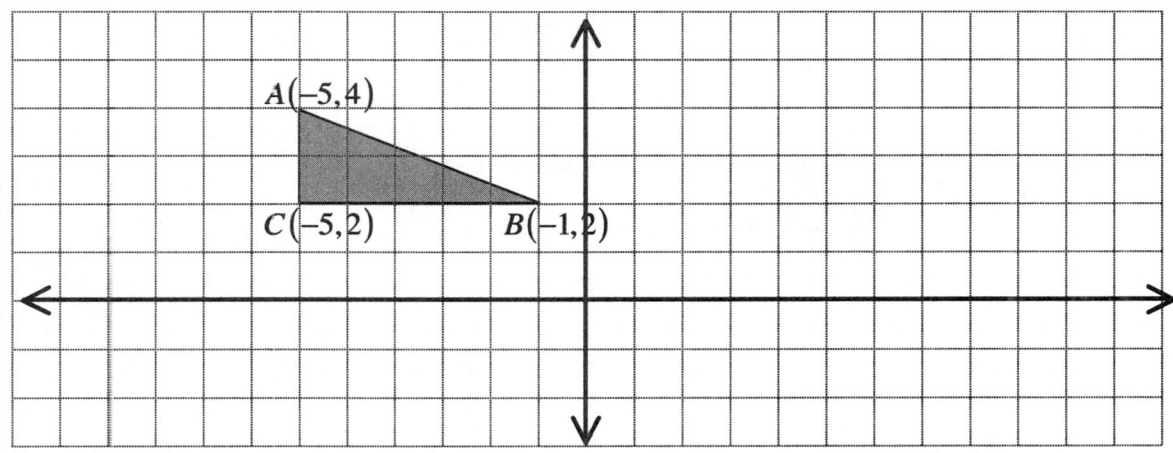

    Determine another transformation that maps $\triangle ABC$ onto $\triangle A"B"C"$. (Use proper transformation notation to express your answer.

18. (a) The composition of a rotation and a reflection through a line that passes through the center of the rotation is a _____.
    (b) The composition of a rotation and a reflection through a line that does not pass through the center of the rotation is a _____.

CHAPTER 6

TRANSFORMATIONAL GEOMETRY
Compositions of Isometries and Dilations, Continued

19. Consider $\triangle ABC$ with coordinates as shown in the figure below.

    Draw and label $\triangle DEF$, the image of $\triangle ABC$, after the transformation $r_{(O,90°)}$.

    Draw and label $\triangle GHI$, the image of $\triangle DEF$, after the transformation $T_{(-4,-6)}$.

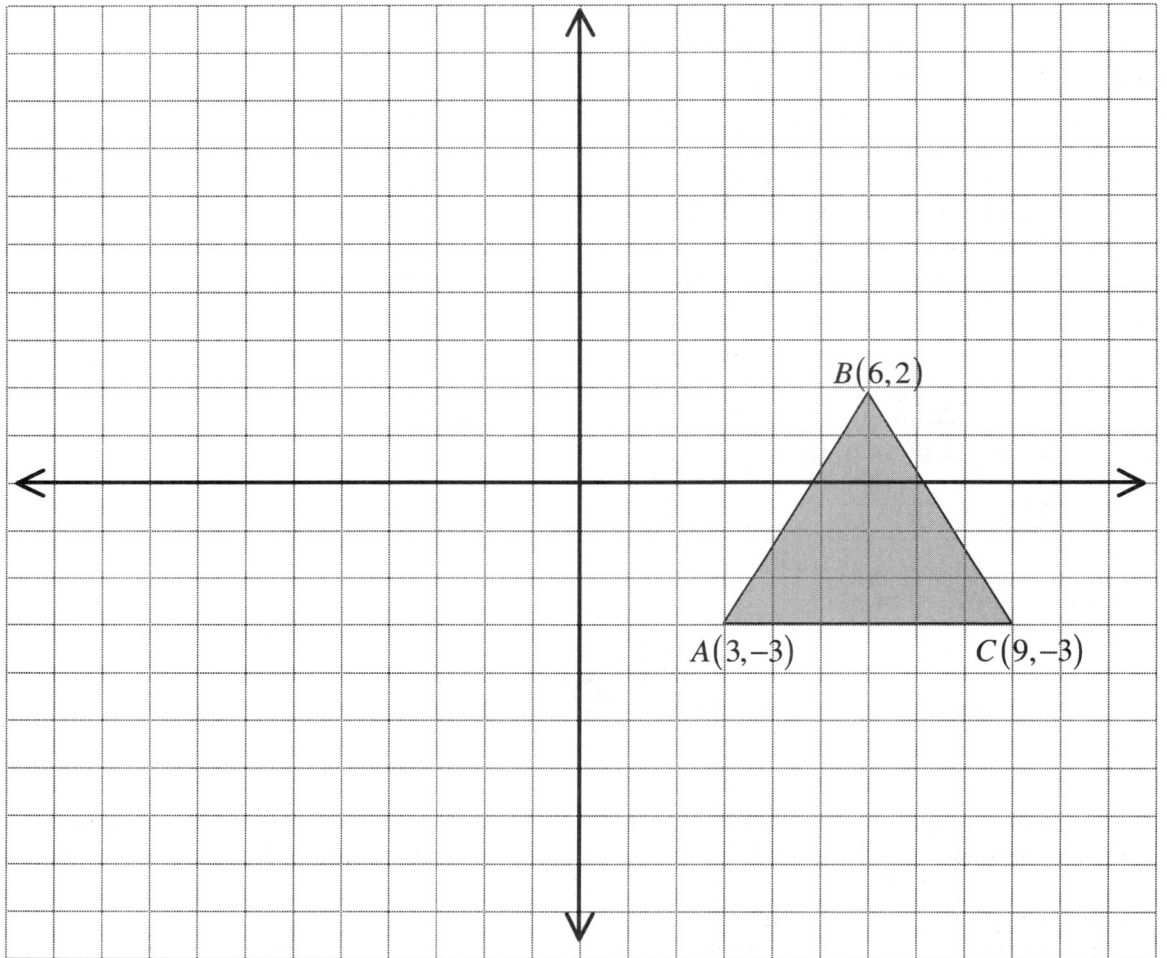

Using your knowledge of slope and perpendicular lines, draw the perpendicular bisectors of $\overline{AG}$, $\overline{BH}$ and $\overline{CI}$. What is the single point of intersection of the three perpendicular bisectors?

How does the point of intersection of the three perpendicular bisectors relate to the single transformation that maps $\triangle ABC$ onto $\triangle GHI$?

Chapter 6

TRANSFORMATIONAL GEOMETRY

Define, Investigate, Justify and Apply Isometries in the Plane

REMEMBER

An **isometry** is a transformation in which the image is congruent to the original figure. This is because the distance between any two points in the original figure is equal to the distance between the corresponding points of the transformed figure.

Examples of isometries are rotations, line reflections, translations and glide reflections.

Isometries can be **direct** (orientation preserved) or **opposite** (orientation switched).

Direct isometries are rotations and translations.
Opposite isometries are line reflections and glide reflections.

Example: In Quadrant I, draw $\overline{AB}$ so that it connects the points (0, 3) to (3, 0). Let $\overline{A'B'}$ be the image of $\overline{AB}$ under a reflection through the origin. What type of quadrilateral is $ABA'B'$? Justify your answer.

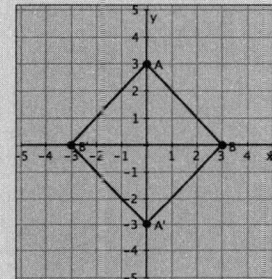

Answer: Draw a picture of $ABA'B'$ to help visualize the figure.

$ABA'B'$ is a square because all the sides are equal in length and each angle measures 90°. (These can be verified using coordinate geometry formulas such as the distance formula and the slope formula.)

---

1. Which of the following is *not* an example of an isometry?

    (1) Rotation (3) Translation
    (2) Dilation (4) Line Reflection

    _____

2. An isometry occurs when a figure and its transformation are

    (1) Congruent (3) Similar
    (2) Equal   (4) A tautology

    _____

3. Draw a line through the figure below to create two figures that form an isometry. Justify your answer.

4. A direct isometry is different from an opposite isometry because a direct isometry preserves

    (1) Area   (3) Orientation
    (2) Distance   (4) Angle Measurements

    _____

5. In Quadrant I, draw $\overline{AB}$ so that it connects the points $A(0,7)$ and $B(x,0)$. Let $\overline{AB'}$ be the reflection of $\overline{AB}$ over the y-axis. Determine the values of x for which $\triangle BAB'$ will be an obtuse triangle?

Chapter 6

TRANSFORMATIONAL GEOMETRY
Define, Investigate, Justify and Apply Isometries in the Plane, (continued)

5. In Quadrant I, draw $\overline{AB}$ so that it is parallel to the x-axis. Let $\overline{A'B'}$ be the reflection of $\overline{AB}$ through the origin. What type of quadrilateral is $ABA'B'$? Justify your answer.

6. In Quadrant I, draw $\overline{AB}$ so that it is parallel to the x-axis. Let $\overline{A'B'}$ be the result after a glide reflection of $\overline{AB}$ down 5 spaces and then reflected over the y-axis. What type of quadrilateral is $ABA'B'$? Justify your answer.

7. In Quadrant I, draw right $\triangle ABO$ so that its vertices are the points $A(0,5)$, $B(2,0)$ and $O(0,0)$. Describe three different transformations of $\triangle ABO$ that are necessary to create the figure drawn below.

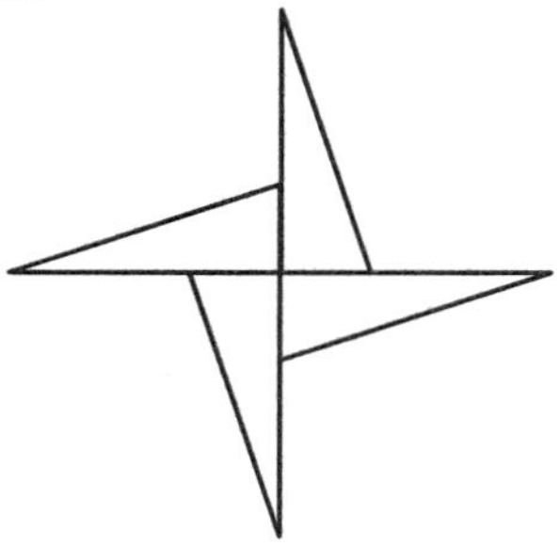

Chapter 6

TRANSFORMATIONAL GEOMETRY
Define, Investigate, Justify and Apply Isometries in the Plane, (continued)

8. In the accompanying diagram, $\overline{PR} \perp \overline{QT}$, $\overline{UQ} \perp \overline{QT}$, $\overline{QR} \cong \overline{TS}$, $\overline{PR} \cong \overline{UQ}$ and $\overline{QRST}$. Use the properties of transformations to justify that $\triangle UQS \cong \triangle PRT$

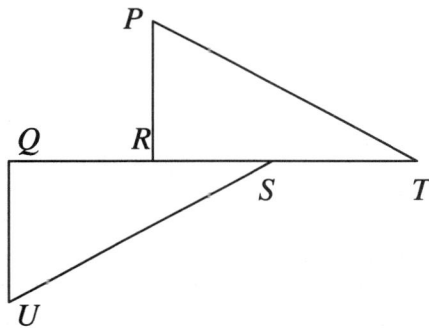

9. Consider $\triangle ABC$ below with point D chosen such that $m\angle ADB = m\angle BDC = m\angle CDA = 120°$. If $\triangle ADC$ is rotated 60° counterclockwise about C, the resulting figure is $\triangle A'D'C$, as shown in the figure below.

Use the properties of transformations to justify that $\overline{A'D'DB}$ is a straight line.

Use the properties of transformations to justify that the length of $\overline{A'B}$ equals the **sum** of the lengths of $\overline{AD}$, $\overline{BD}$ and $\overline{CD}$.

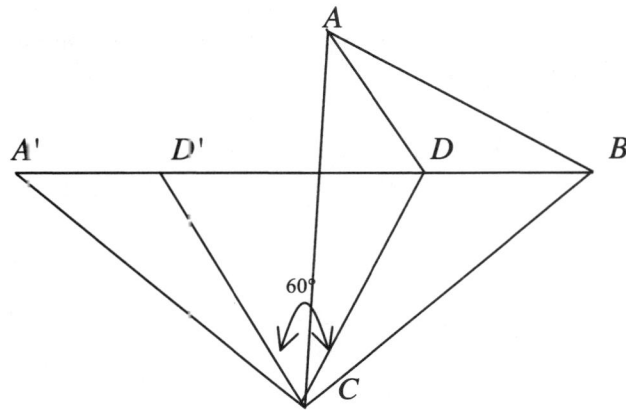

Chapter 6

TRANSFORMATIONAL GEOMETRY
Define, Investigate, Justify and Apply Isometries in the Plane, (continued)

10. In the accompanying diagram, $\overline{AB}$ and $\overline{CD}$ are perpendicular bisectors of each other. Use the properties of transformations to justify that $\triangle ABE \cong \triangle CDE$.

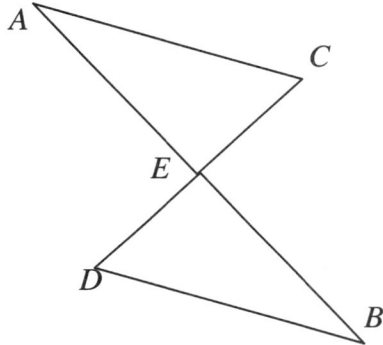

11. Shade part of the figure below that can be used to create the entire figure using transformations. Explain the transformation(s) necessary to create the entire figure from the shaded area.

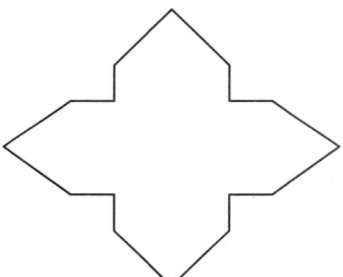

12. In the accompanying diagram of $\triangle ABC$, $\overline{BD}$ is the perpendicular bisector of $\overline{AC}$. Use the properties of transformations to justify that $\triangle ABC$ is isosceles.

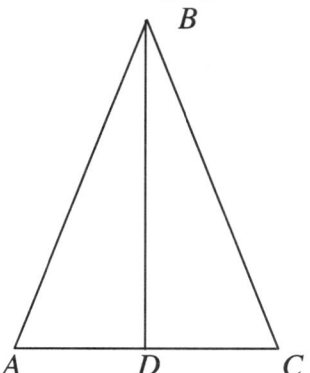

CHAPTER 6

TRANSFORMATIONAL GEOMETRY

Identifying Specific Isometries

REMEMBER

The table below summarizes some properties of figures that are invariant under an isometry.

Property	Line Reflection	Translation	Rotation	Glide Reflections
Angle Measurement	YES	YES	YES	YES
Area	YES	YES	YES	YES
Collinearity	YES	YES	YES	YES
Distance	YES	YES	YES	YES
Orientation	*NO*	*YES*	*YES*	*NO*
Parallelism	YES	YES	YES	YES
Perimeter	YES	YES	YES	YES

Example: In the figure below $\triangle A'B'C'$ is the image of $\triangle ABC$ under an isometry. Using the properties of isometries, determine whether the isometry is a rotation, translation, line reflection or a glide reflection. Explain which properties lead you to your conclusion.

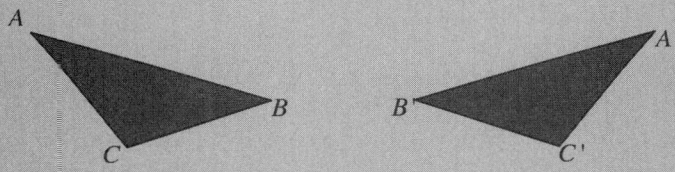

Answer: Since $\triangle ABC$ is oriented clockwise and $\triangle A'B'C'$ is oriented counterclockwise, the isometry is either a line reflection or a glide reflection. Since the perpendicular bisector of segments $\overline{AA'}$, $\overline{BB'}$ and $\overline{CC'}$ are all the same line, the isometry is a line reflection.

1. When inspecting a figure and its image under an isometry, how can it be determined that the isometry is a glide reflection and not a translation?

    (1) Only translations shift an object.
    (2) Only glide reflections switch orientation.
    (3) Only glide reflections change the size of the object.
    (4) Only translations have the segments connecting corresponding vertices parallel to each other.

2. When inspecting a figure and its image under an isometry, how can it be determined that the isometry is a translation and not a rotation?

    (1) Only rotations switch orientation.
    (2) Only rotations change the quadrant of the image and the object.
    (3) Only translations have congruent images and objects.
    (4) Only translations have the segments connecting corresponding vertices parallel to each other.

CHAPTER 6

TRANSFORMATIONAL GEOMETRY
Identifying Specific Isometries, Continued

3. In the figure below quadrilateral $J'U'D'Y$ is the image of quadrilateral $JUDY$ under an isometry. Using the properties of isometries, determine whether the isometry is a rotation, translation, line reflection or a glide reflection. Explain which properties lead you to your conclusion.

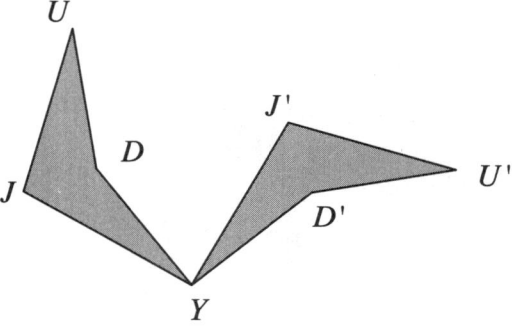

4. In the figure below pentagon $W'H'I'T'E$ is the image of pentagon $WHITE$ under an isometry. Using the properties of isometries, determine whether the isometry is a rotation, translation, line reflection or a glide reflection. Explain which properties lead you to your conclusion.

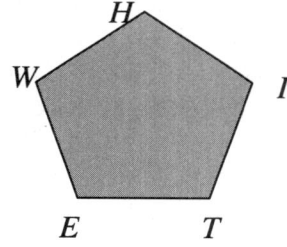

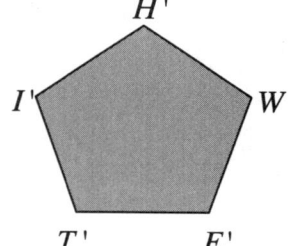

CHAPTER 6

TRANSFORMATIONAL GEOMETRY
Identifying Specific Isometries, Continued

5. In the figure below △A'B'C' is the image of △ABC under an isometry. Using the properties of isometries, determine whether the isometry is a rotation, translation, line reflection or a glide reflection. Explain which properties lead you to your conclusion.

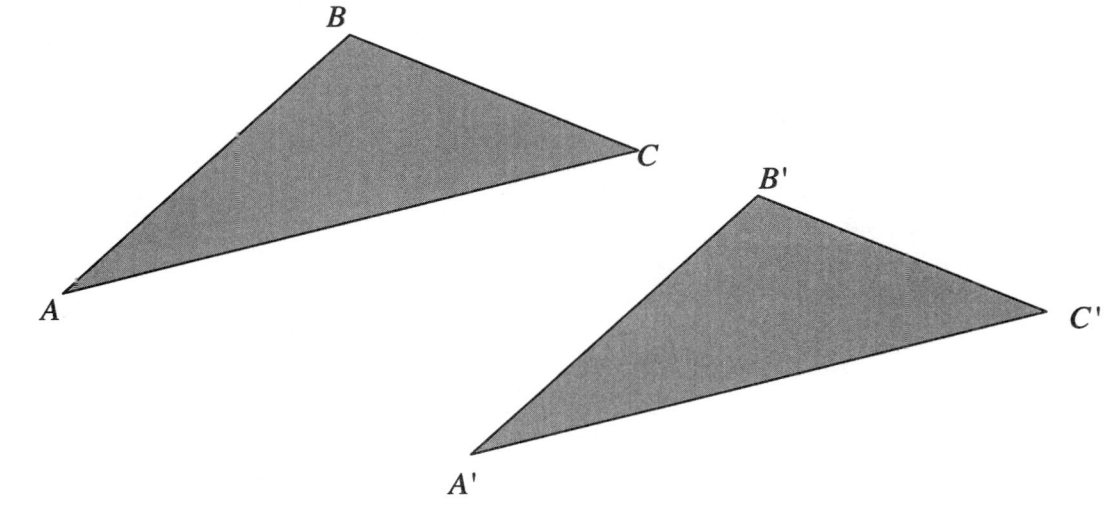

6. In the figure below hexagon M'E'G'H'A'N' is the image of hexagon MEGHAN under an isometry. Using the properties of isometries, determine whether the isometry is a rotation, translation, line reflection or a glide reflection. Explain which properties lead you to your conclusion.

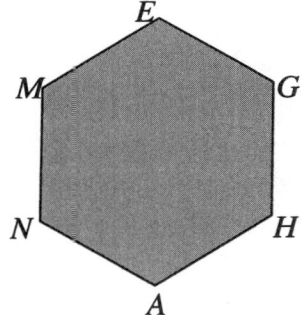 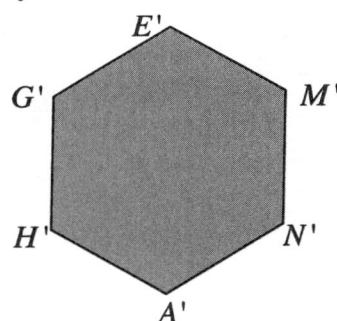

CHAPTER 6

TRANSFORMATIONAL GEOMETRY
Identifying Specific Isometries, Continued

7. In the figure below rhombus $R'O'S'E$ is the image of rhombus $ROSE$ under an isometry. Using the properties of isometries, determine whether the isometry is a rotation, translation, line reflection or a glide reflection. Explain which properties lead you to your conclusion.

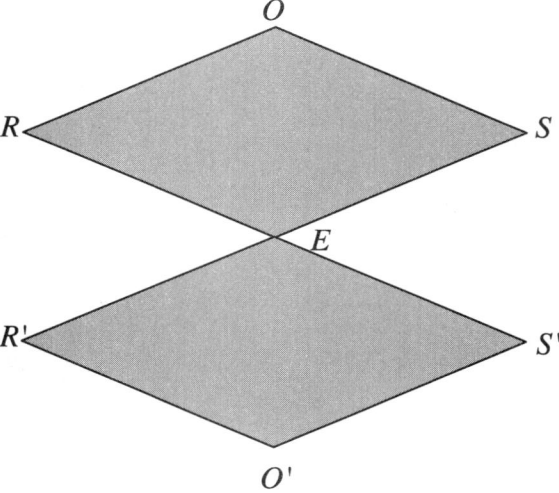

8. In the figure below figure $A'M'B'I'XE'DT'R'O'U'S'$ is the image of figure $AMBIDEXTROUS$ under an isometry. Using the properties of isometries, determine whether the isometry is a rotation, translation, line reflection or a glide reflection. Explain which properties lead you to your conclusion.

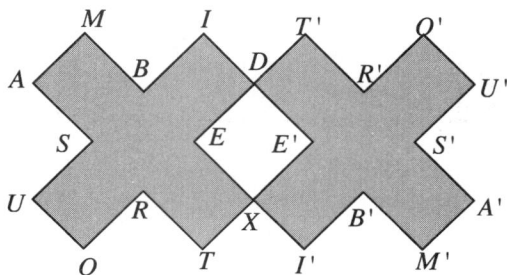

CHAPTER 6

TRANSFORMATIONAL GEOMETRY

Identifying Specific Isometries, Continued

9. On the axes below, ΔB is the image of ΔA after a rotation. Determine the angle of rotation and the center of rotation. Explain the method used to reach your answer.

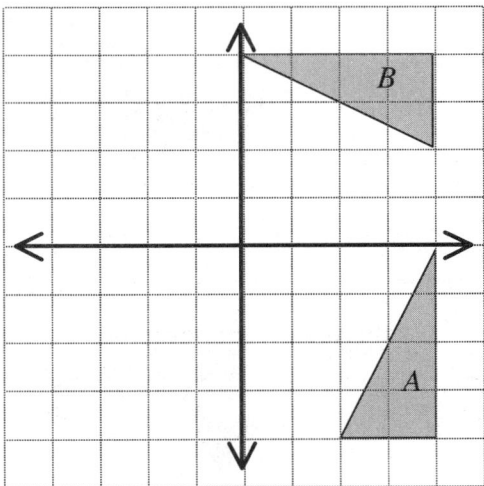

10. On the axes below, ΔB is the image of ΔA after a rotation. Determine the angle of rotation and the center of rotation. Explain the method used to reach your answer.

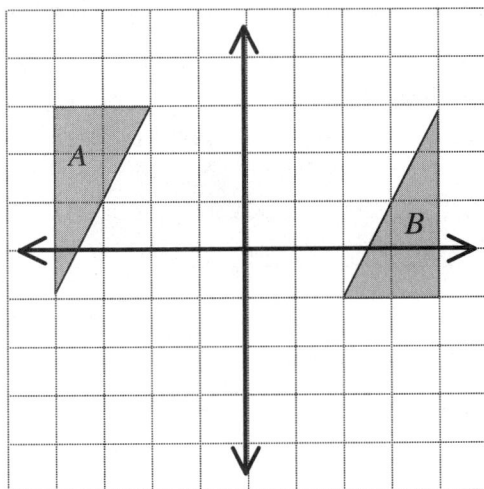

CHAPTER 6

TRANSFORMATIONAL GEOMETRY

Identifying Specific Isometries, Continued

11. On the axes below, $\triangle B$ is the image of $\triangle A$ after a rotation. Determine the angle of rotation and the center of rotation. Explain the method used to reach your answer.

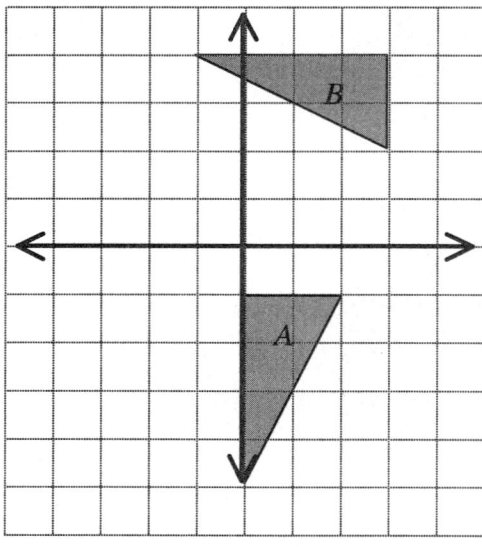

12. On the axes below, equilateral $\triangle B$ is the image of equilateral $\triangle A$ after different transformations. (Express answers using proper notation.)

    Determine the line of reflection that that maps $\triangle A$ onto $\triangle B$.

    Determine an angle of rotation and a center of rotation that maps $\triangle A$ onto $\triangle B$.

    Determine the translation and vector of translation that maps $\triangle A$ onto $\triangle B$.

    Aidan conjectures that line ℓ can also be used to map $\triangle A$ onto $\triangle B$. What type of transformation is Aidan considering? Is he correct? Explain which properties of this transformation lead you to your conclusion.

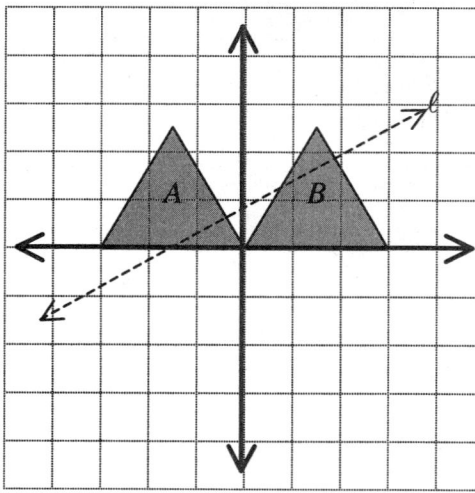

132

CHAPTER 6

TRANSFORMATIONAL GEOMETRY
Identifying Specific Isometries, Continued

13. On the axes below, rhombus B is the image of rhombus A after different transformations. (Express answers using proper notation.)

    Determine the line of reflection that that maps rhombus B onto rhombus A.

    Determine an angle of rotation and a center of rotation that maps rhombus B onto rhombus A.

    Determine the translation and vector of translation that maps rhombus B onto rhombus A.

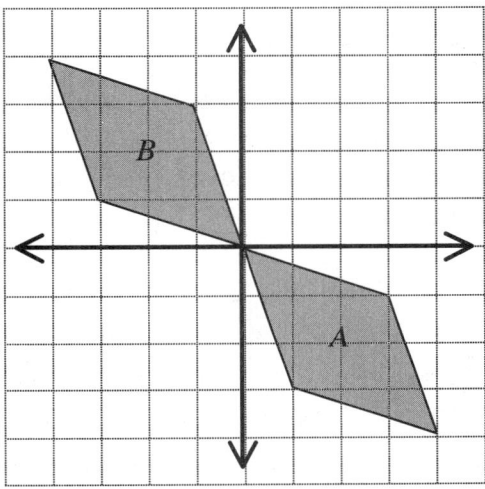

14. Name isometries which will map figure A onto figures B, C, D, E, F, G and H.

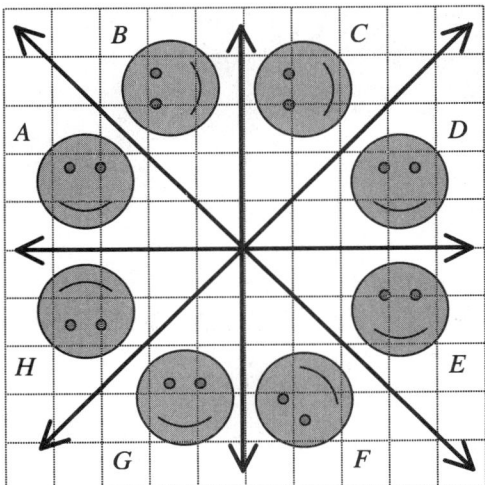

133

CHAPTER 6

TRANSFORMATIONAL GEOMETRY
Identifying Specific Isometries, Continued

15. On the axes below, square B is the image of square A after different transformations. (Express answers using proper notation.)

    Determine a glide reflection that maps square A onto square B.

    Is square B the image of square A after $R_{(Origin, 180°)}$? Justify your answer using the properties of isometries.

    Determine the translation and vector of translation that maps square B onto square A.

    Linda conjectures that square B is the image of square A after a 90° counterclockwise rotation about the point $(1, 3)$. Explain whether Linda's conjecture is correct or incorrect. If her conjecture is incorrect, explain how her conjecture can be corrected.

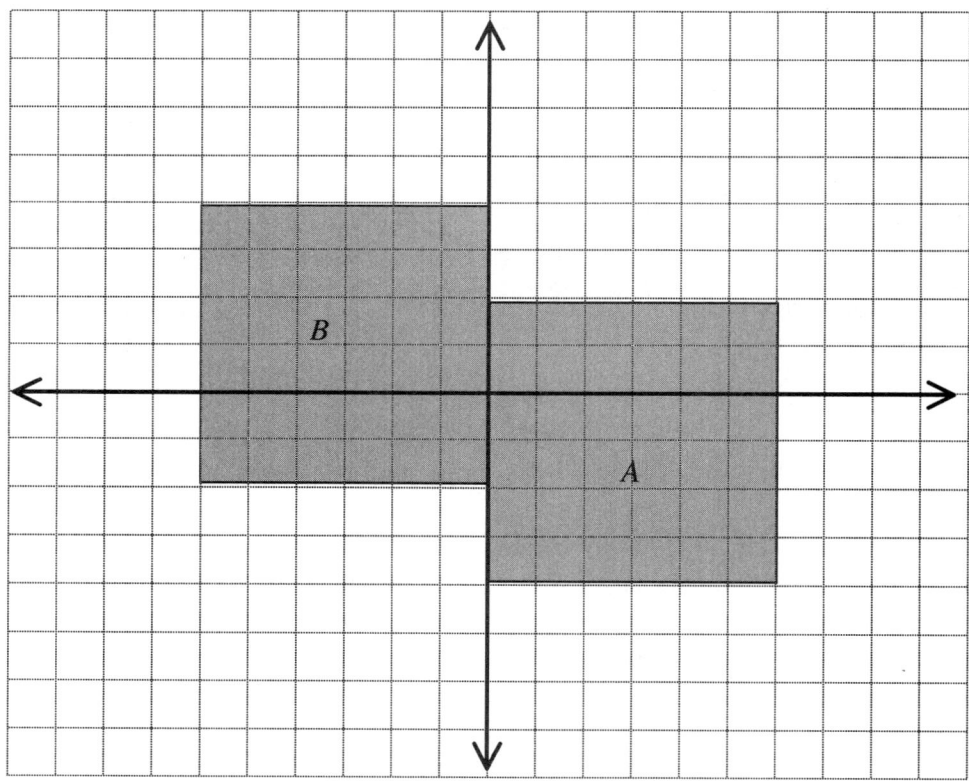

CHAPTER 6

TRANSFORMATIONAL GEOMETRY
Identifying Specific Isometries, Continued

16. In the diagram below, figure $ABCD$ is a rhombus and $\overline{AC}$ and $\overline{BD}$ are diagonals that intersect at point E. Identify, using correct notation, isometries which can be used to map ΔI onto ΔII, ΔIII and ΔIV.

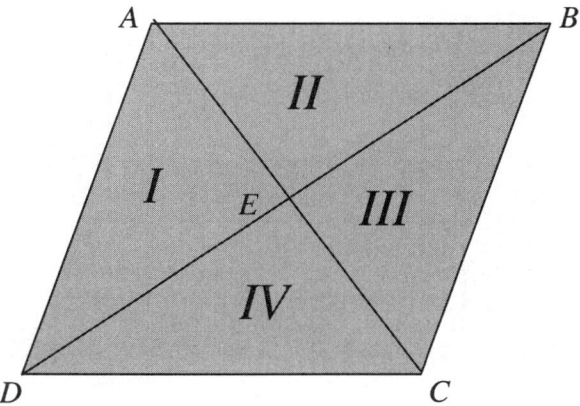

17. In the diagram below, figure ISO is an isosceles right triangle with right angle S, $\overline{SE} \cong \overline{SF}$ and $\overline{OE}$ and $\overline{IF}$ intersect at G. Identify, using correct notation, which isometries can be used to map ΔISF onto ΔOSE. Name a triangle congruence theorem that could be used to prove that $\Delta ISF \cong \Delta OSE$. Justify your answer.

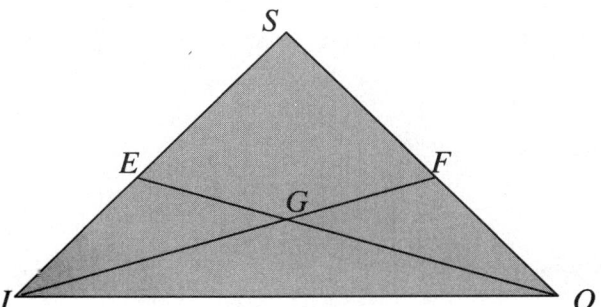

135

CHAPTER 6

TRANSFORMATIONAL GEOMETRY
Identifying Specific Isometries, Continued

18. In the diagram below, figure $ABCD$ is a trapezoid with $\overline{AB} \parallel \overline{DC}$, $\overline{AD} \perp \overline{AB}$, $\overline{AD} \perp \overline{DC}$, $\overline{AE} \cong \overline{DC}$ and $\overline{AB} \cong \overline{DE}$.

    Identify, using correct notation, the isometries necessary to map $\triangle EAB$ onto $\triangle CDE$.

    Justify that $\triangle BEC$ is an isosceles right triangle.

    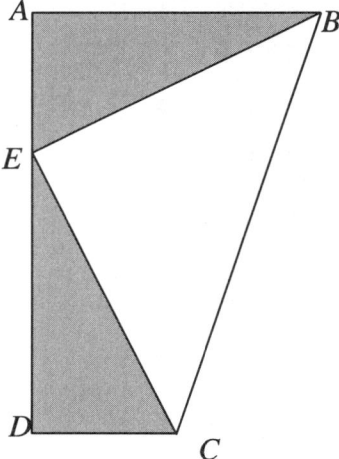

19. In the diagram below, figures ABC and DBE are both isosceles triangles with $\overline{AB} \cong \overline{CB}$ and $\overline{DB} \cong \overline{EB}$. If $m\angle BAD = 48°$ and $m\angle ADB = 104°$, identify, using proper notation, the isometries necessary to map $\triangle ABC$ onto $\triangle CBE$.

    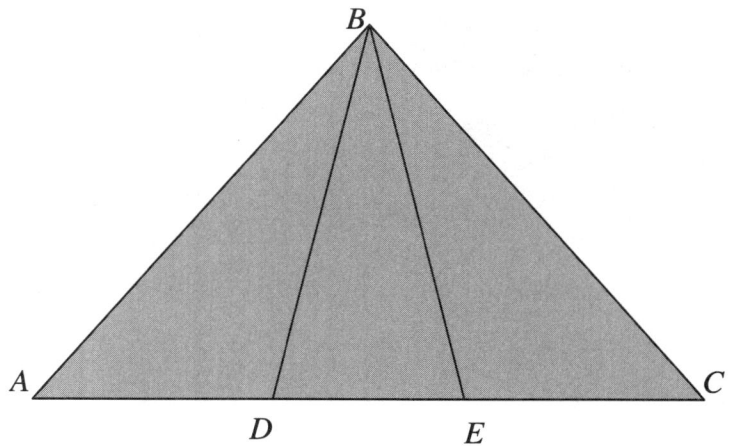

CHAPTER 6

TRANSFORMATIONAL GEOMETRY
Identifying Specific Isometries, Continued

20. In figure $ABCDEFGHIJKL$ below, $\angle C$, $\angle E$, $\angle I$ and $\angle K$ are right angles, $\overline{AM} \cong \overline{GH}$ and $\overline{MC} \cong \overline{HE} \cong \overline{MK} \cong \overline{HI}$. $\triangle AEH$, $\triangle GKM$ and $\triangle AIH$ are images of $\triangle GCM$ under different isometries.

    Identify, using proper notation, the isometries necessary to map $\triangle GCM$ onto $\triangle AEH$, $\triangle GKM$ and $\triangle AIH$.

    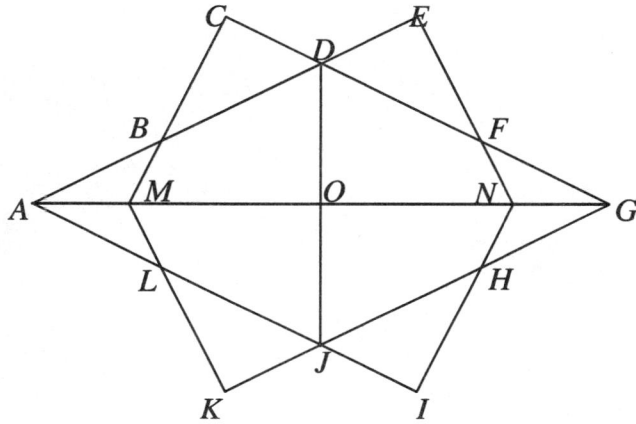

21. In the diagram below, the unshaded triangle is the image of the shaded triangle under different isometries.

    Draw the line of reflection that maps the shaded triangle onto the unshaded triangle. Label the line ℓ.

    Draw a center of a rotation, labeled C, that can be used to map the shaded triangle onto the unshaded triangle. Estimate the angle of rotation. Justify your answer.

    Draw a line of reflection and a vector of translation that maps the shaded triangle onto the unshaded triangle under a glide reflection. Label the line k.

    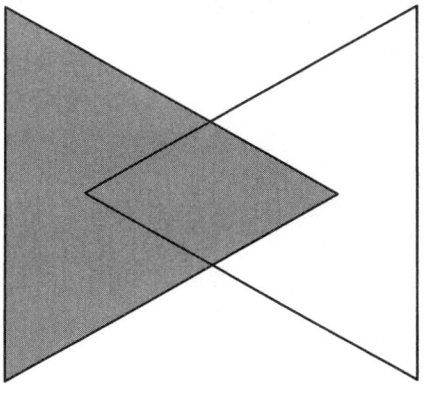

137

CHAPTER 6

TRANSFORMATIONS

Identify Specific Similarities

> **REMEMBER**
>
> When investigating a transformation of an object that contains a dilation, consider the object and its image's orientation, invariant points and any parallel lines that occur.
>
> Under a dilation, each line or line segment will be transformed to a line or line segment that is parallel to the original line or line segment.
>
> Triangles can be shown to be similar, and thus a dilation of each other, using any of the following three theorems:
>
> | Theorem | What the theorem means. |
> |---------|-------------------------|
> | A.A.A. | 3 pairs of corresponding angles are congruent. |
> | S.A.S. | 2 pairs of corresponding sides are in proportion and the included angles are congruent. |
> | S.S.S. | 3 pairs of corresponding sides are in proportion. |

Example: What isometry maps $\triangle ABC$ to $\triangle A'BC'$?

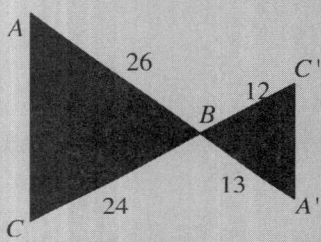

Answer: Since the measure of $\overline{AB}$ is twice the measure of $\overline{A'B}$, the measure of $\overline{BC}$ is twice the measure of $\overline{BC'}$, and $\angle ABC \cong \angle A'BC'$, $\triangle ABC \approx \triangle A'BC'$.

The isometry that maps $\triangle ABC$ to $\triangle A'BC'$ is $R_{(B,180°)} \circ D_{\frac{1}{2}}$ or $D_{\frac{1}{2}} \circ R_{(B,180°)}$.

1. Which of the following is equivalent to a dilation with a negative scale factor?

   (1) dilation and rotation
   (2) dilation and translation
   (3) dilation and line reflection
   (4) dilation and glide reflection

   _____

2. Line segments connecting corresponding vertices of two similar objects

   (1) meet on a circle
   (2) meet on a line
   (3) meet at a point
   (4) are parallel

   _____

3. Which of the following is true about concentric figures?

   (1) They are dilations of each other with a center of dilation outside the figures.
   (2) They are dilations of each other with a center of dilation inside the figures.
   (3) They are dilations of each other with the center of dilation at a vertex of one of the figures.
   (4) They are not dilations of each other.

   _____

CHAPTER 6

TRANSFORMATIONS

Identify Specific Similarities, Continued

4. In the figure below, D is the midpoint of $\overline{AB}$ and E is the midpoint of $\overline{AC}$. Kara conjectures that if she connects the midpoints of two sides of $\triangle ABC$, she forms $\triangle ADE$ which is a dilation of $\triangle ABC$.

   Explain whether Patricia's conjecture is correct.

   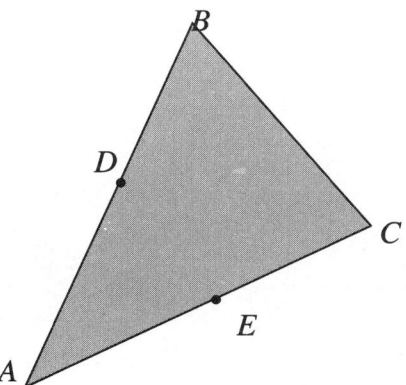

5. Consider $\triangle ABC$ below with A', the midpoint of $\overline{BC}$, B', the midpoint of $\overline{AC}$ and C', the midpoint of $\overline{AB}$. Let D be the circumcenter, E be the centroid and F be the orthocenter of $\triangle ABC$.

   Identify, using proper notation, the transformation(s) necessary to map $\triangle ABC$ onto $\triangle A'B'C'$.

   Which point, D, E or F represents the orthocenter of $\triangle A'B'C'$? Justify your answer using the properties of dilations.

   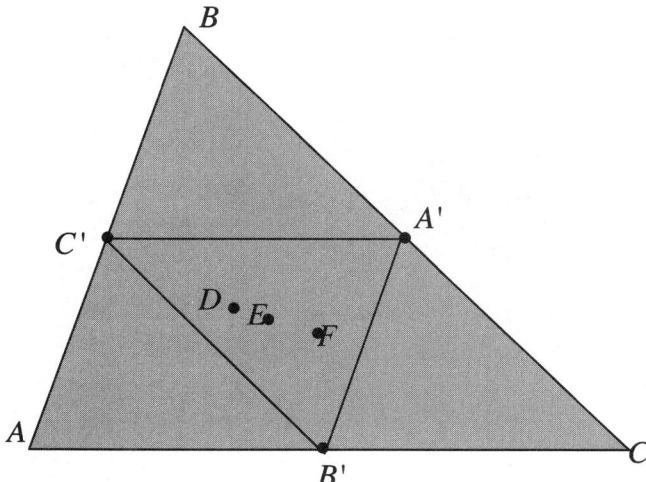

139

CHAPTER 6

TRANSFORMATIONS

Identify Specific Similarities, Continued

6. In the figure below, $\triangle ABC$ is a right triangle with right angle B, $\overline{BD} \perp \overline{AC}$ and $AB = 6$, $AD = 3$ and $CD = 9$.

   Determine the transformations, using proper notation, necessary to map $\triangle ABD$ onto $\triangle BCD$.

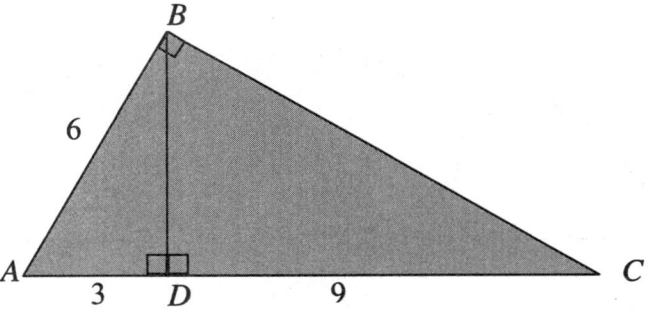

7. In the figure below, $\triangle ABC$ is a right triangle with right angle B, $\overline{BD} \perp \overline{AC}$ and $AB = 6$, $AD = 3$ and $CD = 9$.

   Determine the transformations, using proper notation, necessary to map $\triangle ABD$ onto $\triangle ACB$.

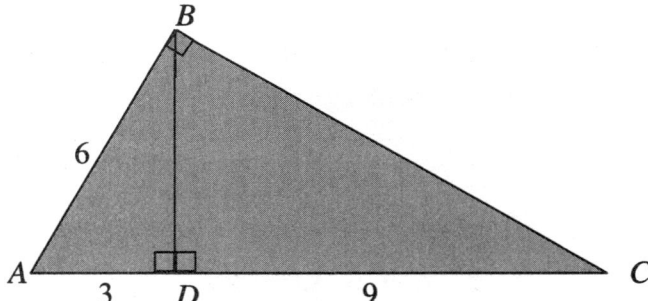

CHAPTER 6

TRANSFORMATIONS

Identify Specific Similarities, Continued

8. Consider the three ellipses graphed below.

    Write a composition of transformations that will map ellipse A onto ellipse B.

    Write a composition of transformations that will map ellipse A onto ellipse C.

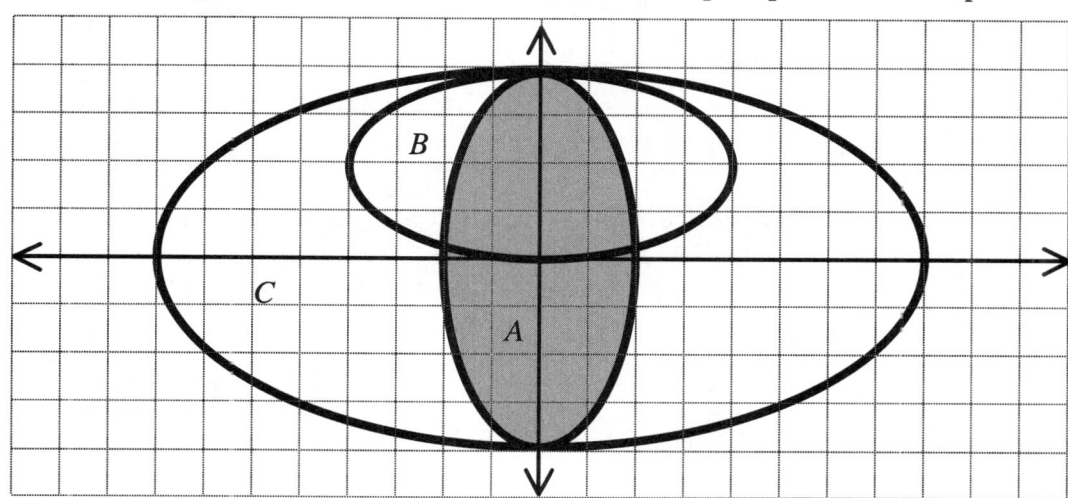

9. The equation of the shaded circle below is $(x+1)^2 + y^2 = 1$.

    Write, using proper notation, a transformation that will map $(x+1)^2 + y^2 = 1$ onto $(x+3)^2 + y^2 = 1$.

    Write, using proper notation, a composition of transformations that will map $(x+1)^2 + y^2 = 1$ onto $(x-2)^2 + y^2 = 4$.

    Write, using proper notation, a composition of transformations that will map $(x+1)^2 + y^2 = 1$ onto $x^2 + y^2 = 16$.

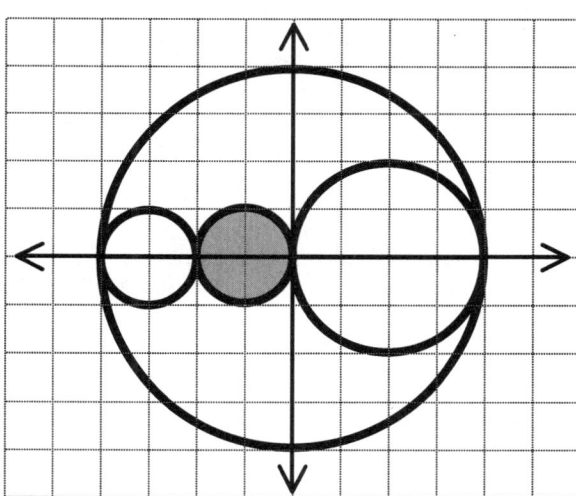

Chapter 6

TRANSFORMATIONAL GEOMETRY

Properties that Remain Invariant Under Isometries and Similarities

> **REMEMBER**
>
> An **isometry** is a transformation in a plane that results in an image that is *congruent* to the original object. The four isometries in the plane are **reflections, translations, rotations** and **glide reflections**.
>
> The composition of two or more isometries is also an isometry. The composition of two isometries is often *not commutative*, although it can be.
>
> Any object in a plane can be mapped to a congruent object using at most three isometries.
>
> The table below summarizes some properties of figures that are invariant under an isometry.
>
> | Property | Line Reflection | Translation | Rotation | Glide Reflections |
> |---|---|---|---|---|
> | Angle Measurement | YES | YES | YES | YES |
> | Area | YES | YES | YES | YES |
> | Collinearity | YES | YES | YES | YES |
> | Distance | YES | YES | YES | YES |
> | *Orientation* | *NO* | YES | YES | *NO* |
> | Parallelism | YES | YES | YES | YES |
> | Perimeter | YES | YES | YES | YES |
>
> **Example:** A glide reflection consists of two parts; a reflection and a translation. Which of these causes the orientation of an object to switch?
>
> **Answer:** The reflection.

1. A direct isometry is one that preserves orientation. Which two of the four isometries in a plane are direct isometries?

2. An opposite isometry is one that does not preserve orientation. Which two of the four isometries in a plane are opposite isometries?

3. Which point in the plane is invariant under a rotation?

4. Which type of isometry is the composition of two reflections?

    (1) Direct   (3) Neither
    (2) Opposite (4) Could be either

    _____

5. Name a translation that keeps all points in the plane invariant.

6. Under a line reflection, which point(s) remain invariant?

Chapter 6

TRANSFORMATIONAL GEOMETRY
Properties that Remain Invariant Under Isometries and Similarities, (continued)

7. Nancy believes that the reflection of $\triangle ABC$ through two parallel lines that are 17 inches apart is equivalent to a translation. Explain why Nancy is correct using your knowledge of isometries. Describe the translation using proper notation. 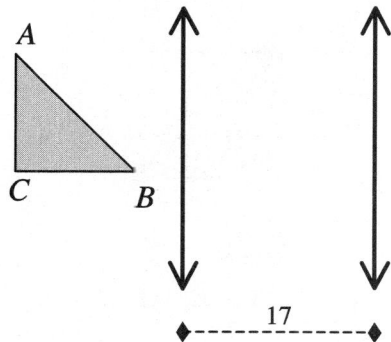	9. Patricia conjectures that a translation is equivalent to two reflections over different lines. Check her conjecture by drawing two lines that can be used as lines of reflection to map quadrilateral $JACK$ to quadrilateral $J'A'C'K'$. 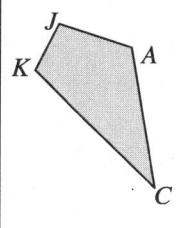 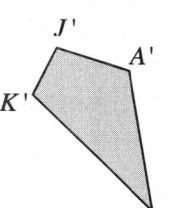 Use the properties of isometries to justify why your reflections are equivalent to a translation.
8. Annmarie believes that the reflection of $\triangle DEF$ through two intersecting lines $50°$ apart is equivalent to a rotation. Check Annmarie by reflecting $\triangle DEF$ over line ℓ and then line m. Explain why Annmarie is correct using the properties of isometries. 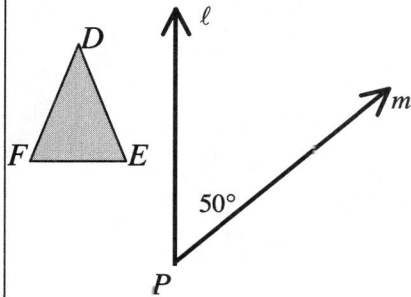 Determine the angle of rotation.	10. In the figure below, $\overline{FG}$ is the perpendicular bisector of $\overline{AB}$ and $\overline{CD}$. Frank believes that $\triangle BHC$ is the reflection of $\triangle AHD$ over $\overline{FG}$. Using your knowledge of isometries, explain whether Frank is correct. 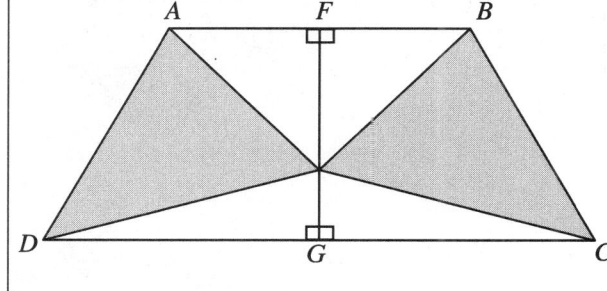

143

Chapter 6

TRANSFORMATIONAL GEOMETRY
Properties that Remain Invariant Under Isometries and Similarities, (continued)

11. Graph parallelogram $JOHN$ where $J(-3,4)$, $O(-2,1)$, $H(-5,-5)$ and $N(-6,-2)$.

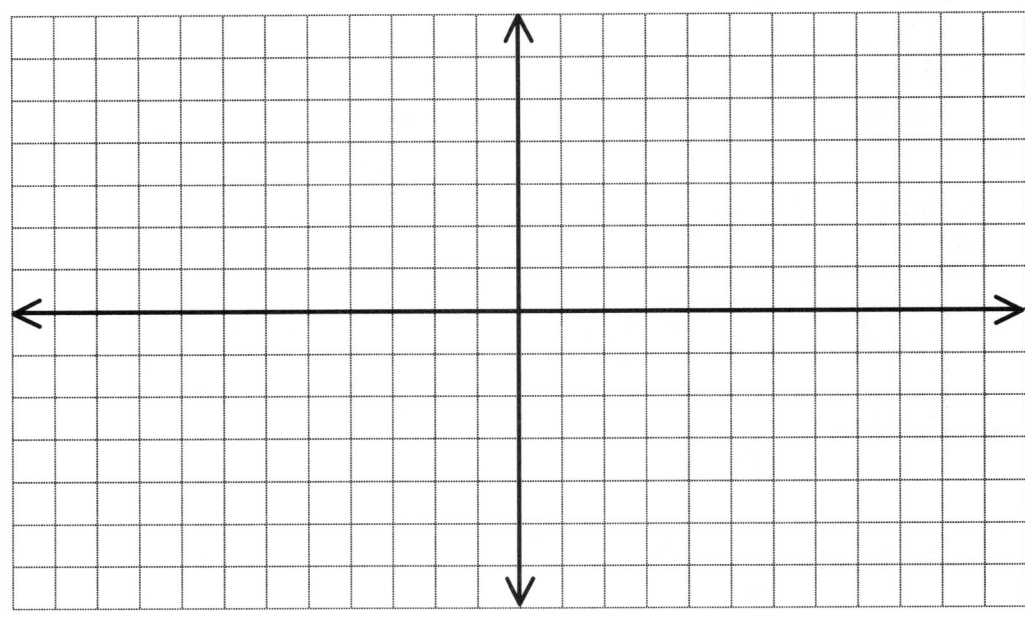

$\overline{JN}$ is part of the line whose equation is $y = 2x + 10$. Use $y = 2x + 10$ to show that the point $(-4, 2)$ is on $\overline{JN}$.

Graph $J'O'H'N'$, the image of $JOHN$ after the transformation $R_{(Origin,-90°)}$.

Use coordinate geometry to justify that $J'O'H'N'$ is a parallelogram.

Determine the equation of the line passing through the points J' and N' and use it to show that the transformation of the point $(-4, 2)$ is on $\overline{J'N'}$.

<u>Fill in the blanks in the following statement:</u>
This problem helps demonstrate that the properties _____ and

_____ are _____ under a _____.

Chapter 6

TRANSFORMATIONAL GEOMETRY
Properties that Remain Invariant under Isometries and Similarities, (continued)

12. Graph $\triangle TRY$ with coordinates $T(1,-2)$, $R(3,-6)$ and $Y(-3,-6)$.

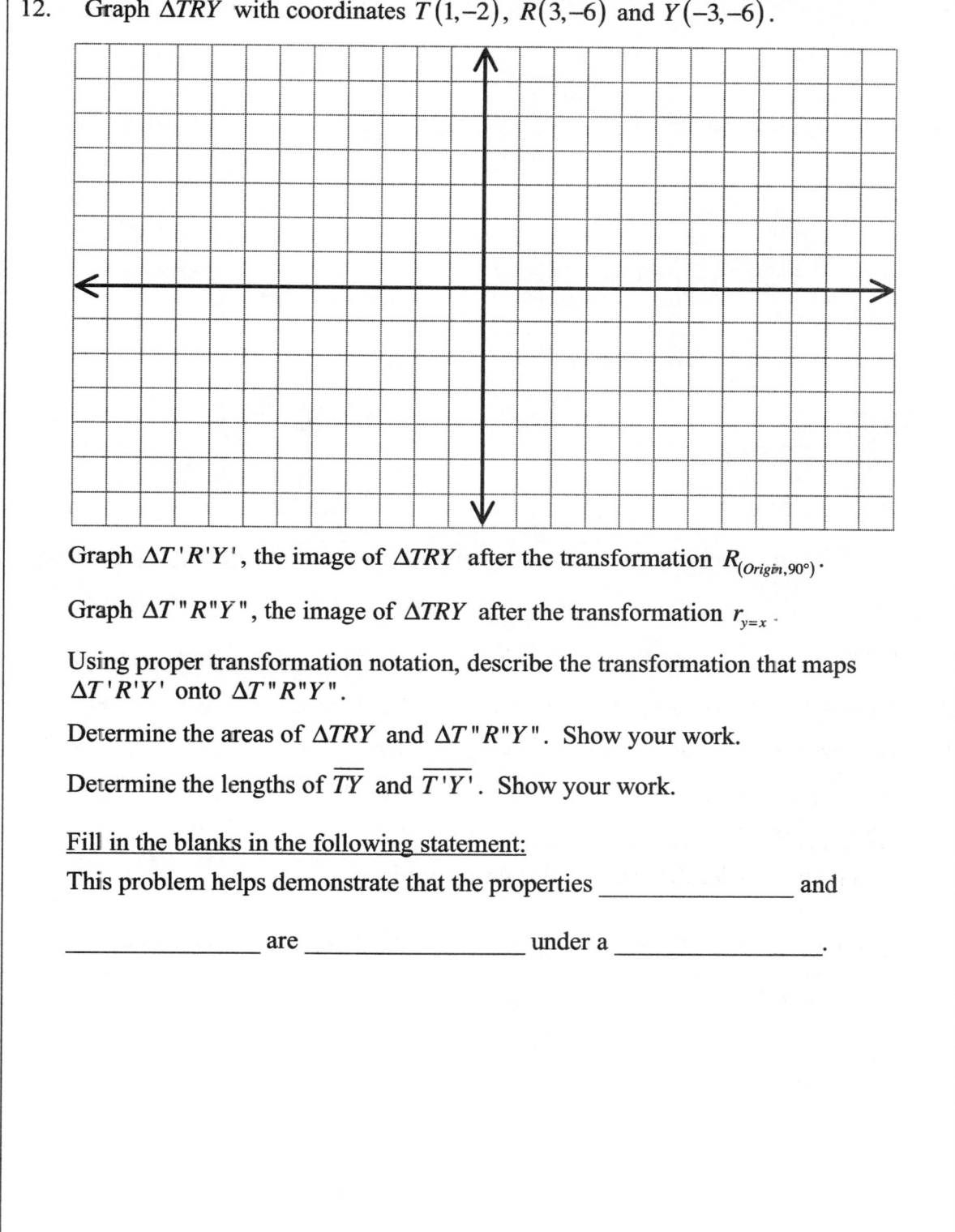

Graph $\triangle T'R'Y'$, the image of $\triangle TRY$ after the transformation $R_{(Origin, 90°)}$.

Graph $\triangle T"R"Y"$, the image of $\triangle TRY$ after the transformation $r_{y=x}$.

Using proper transformation notation, describe the transformation that maps $\triangle T'R'Y'$ onto $\triangle T"R"Y"$.

Determine the areas of $\triangle TRY$ and $\triangle T"R"Y"$. Show your work.

Determine the lengths of $\overline{TY}$ and $\overline{T'Y'}$. Show your work.

<u>Fill in the blanks in the following statement:</u>
This problem helps demonstrate that the properties _____ and

_____ are _____ under a _____.

Chapter 6

TRANSFORMATIONAL GEOMETRY
Properties that Remain Invariant Under Similarities

> **REMEMBER**
>
> A **dilation** is a **transformation** in a plane that results in an image that is *similar* to the original object. A **dilation** is a **transformation**, but **_not_** an **isometry**.
>
> The table below summarizes some properties of figures that are invariant under a dilation.
>
> | Property | Dilation | Ratio |
> |---|---|---|
> | Angle Measurement | YES | = |
> | *Area* | NO | $Ratio^2$ |
> | Collinearity | YES | = |
> | *Distance* | NO | Ratio |
> | Orientation | YES | |
> | Parallelism | YES | |
> | *Perimeter* | NO | Ratio |
>
> Triangles can be shown to be similar, and thus a dilation of each other, using any of the following three theorems:
>
> | Theorem | What the theorem means. |
> |---|---|
> | A.A.A. | 3 pairs of corresponding angles are congruent. |
> | S.A.S. | 2 pairs of corresponding sides are in proportion and the included angles are congruent. |
> | S.S.S. | 3 pairs of corresponding sides are in proportion. |
>
> **Example:** Which point is the image of point D after a transformation of D_2 about A?
>
>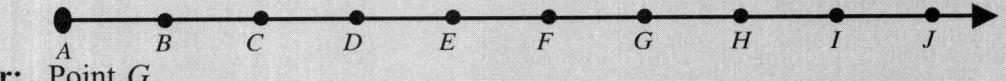
>
> **Answer:** Point G

1. Under a dilation, which point in the plane is invariant?

2. If the area of an object is 6 which of the following is the area of its' image under a dilation of D_3?

   (1) 6   (3) 18
   (2) 2   (4) 54

3. In the figure below, $\triangle ABC \cong \triangle DEF$. Is there a center of dilation? If so, draw its location on the figure. If not, explain why there is none.

   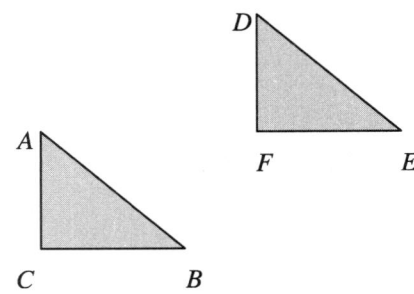

Chapter 6

TRANSFORMATIONAL GEOMETRY
Properties that Remain Invariant Under Similarities, (continued)

4. Graph parallelogram $NICK$ with vertices $N(-2,1)$, $I(2,2)$, $C(4,-1)$ and $K(0,-2)$.

Draw and label $N'I'C'K'$, the image of $NICK$ after a dilation of D_3 with center the origin.

Using coordinate geometry what type of quadrilateral is $N'I'C'K'$?

Determine the areas of $NICK$ and $N'I'C'K'$.

<u>Fill in the blanks in the following statement:</u>

This problem helps demonstrate that _____ is invariant under a

_____ and the _____ of a dilated object is equal to the

_____ of the original object multiple by the scale factor _____.

Chapter 6

TRANSFORMATIONAL GEOMETRY
Properties that Remain Invariant Under Similarities, (continued)

5. Graph isosceles right $\triangle SJV$ with vertices $S(-1,-1)$. $J(1,2)$ and $V(4,0)$.

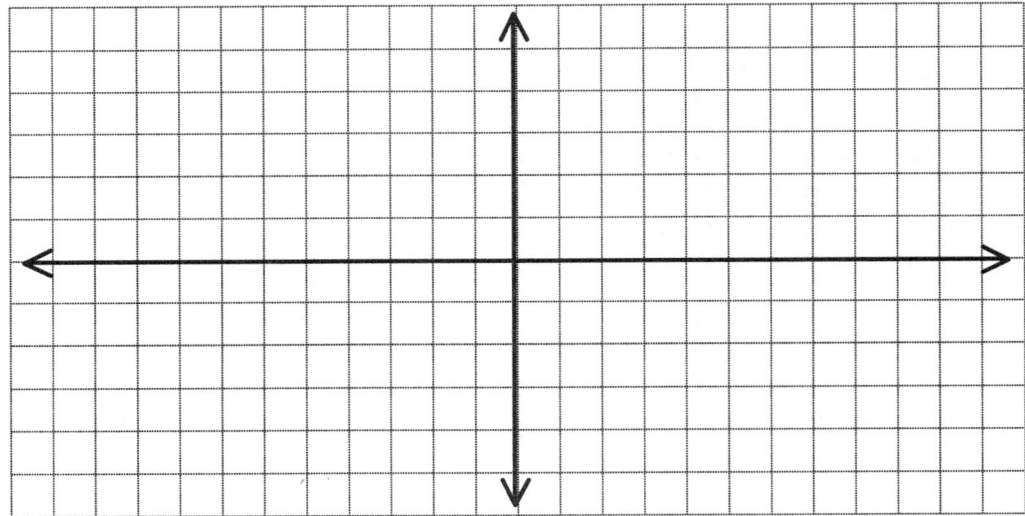

Draw and label $\triangle S'J'V'$, the image of $\triangle SJV$ after a dilation of D_{-2} with center at the origin.

Use coordinate geometry to show that $\triangle S'J'V'$ is also an isosceles right triangle.

What are the measures of $\angle JSV$ and $\angle J'S'V'$?

Using proper notation, express the dilation D_{-2} as the composition of two transformations.

<u>Fill in the blanks in the following statement:</u>

This problem helps demonstrate that _____ is invariant under a

_____ and that the image of a dilated object is _____ to the

original object.

148

Chapter 7
LOCUS

Locus

REMEMBER

A locus is a path of points that satisfy a certain condition. Below are listed fundamental loci theorems that you should be familiar with.

1. The locus of points at a fixed distance, d, from point P, is a circle with the given point P as its center and d as its radius. 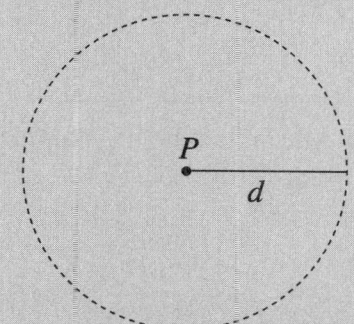	2. The locus of points at a fixed distance, d, from a line ℓ, is a pair of parallel lines d distance from ℓ and on either side of ℓ. 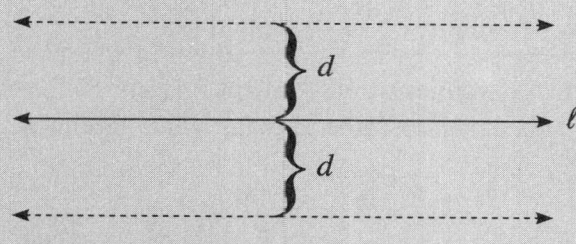
3. The locus of points equidistant from two points, P and Q, is the perpendicular bisector of the line segment determined by the two points. 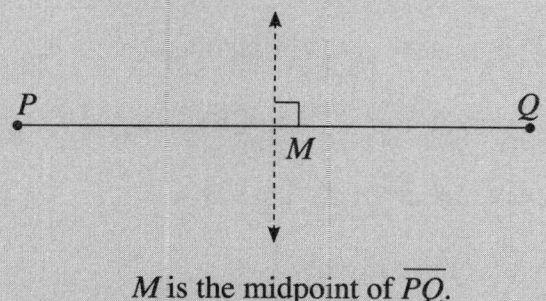 M is the midpoint of $\overline{PQ}$.	4. The locus of points equidistant from two parallel lines ℓ_1 and ℓ_2, is a line parallel to both ℓ_1 and ℓ_2 and midway between them. 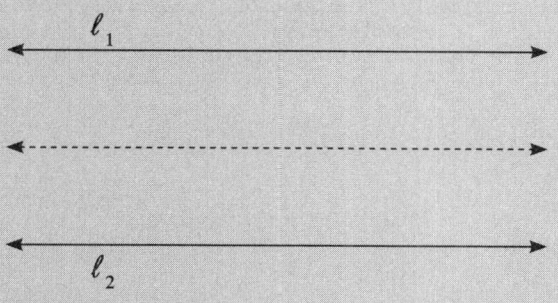

5. The locus of points equidistant from two intersecting lines, ℓ_1 and ℓ_2, is a pair of angle bisectors that bisect the angles formed by ℓ_1 and ℓ_2.

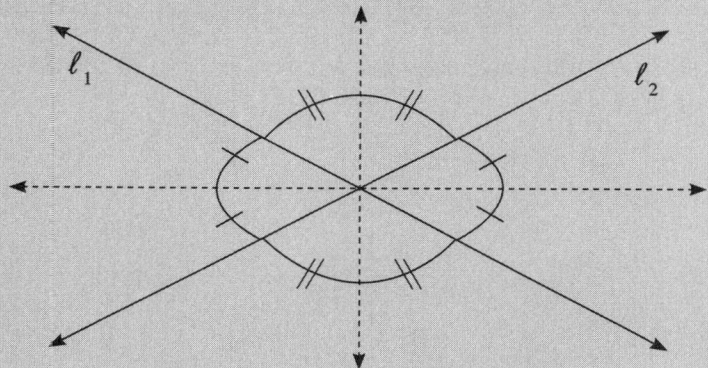

Chapter 7

LOCUS

Locus (continued)

1. Write an equation of the locus of points equidistant from the graphs of the equations of x = 4 and x = –2.

2. A gardener wants to plant flowers 10 feet from the base of a flag pole. When he has completed the task, describe in detail what geometric shape the planted flowers will form.

3. The locus of points equidistant from the sides $\overline{AB}$ and $\overline{AC}$ in scalene $\triangle ABC$ described below is:

   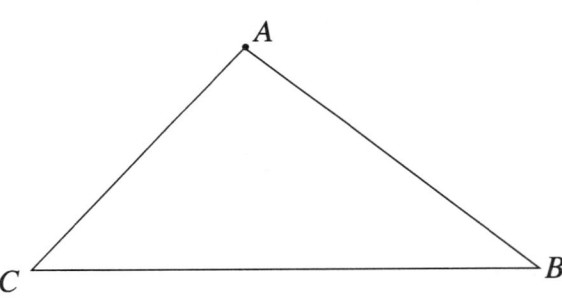

   (1) a medium from $\angle A$ to $\overline{CB}$

   (2) an altitude from $\angle A$ to $\overline{CB}$

   (3) an angel bisector of $\angle A$

   (4) none of the above

4. Describe the locus of points 5 units from the line whose equation of y = 3 on a coordinate graph. Write the equation(s) of the locus.

Chapter 7
LOCUS

Locus (continued)

5. Describe fully the locus of points 4 units from the origin on a coordinate graph. Write an equation of the locus.	9. Describe fully the locus of points equidistant from two concentric circles with radii of 3 and 5.
6. Point A lies on $\overline{PQ}$. The locus of the centers of all circles of radius 4 which pass through point A is (1) one circle (3) two circles (2) two parallel lines (4) one line	10. If the measure of one of the angles formed by two given intersecting lines is 80°, what is the measure of the angle between the two lines that represent the locus of points equidistant from the two given lines? (1) 80° (3) 100° (2) 90° (4) 120°
7. Which of the following is an equation of the locus of points equidistant from the graphs of the equations $x = 6$ and $x = -2$? (1) y = 2 (3) y = 4 (2) x = 2 (4) x = 4	11. What is the slope of the line that is the locus of points equidistant from the points $(1, 7)$ and $(5, -5)$? (1) -3 (3) 3 (2) $-\dfrac{1}{3}$ (4) $\dfrac{1}{3}$
8. Which two points are the locus of points that are equidistant from the equation $y = 5$? (1) (0, 0) and (0, 10) (3) (5, 0) and (0, 5) (2) (0, 3) and (0, 2) (4) (5, 5) and (−5, 5)	12. Which of the following lines is the equation for the locus of points equidistant from the points $(-2, -1)$ and $(2, 5)$? (1) $y = x + 2$ (3) $y = -x + 2$ (2) $y = \dfrac{3}{2}x + 2$ (4) $y = -\dfrac{2}{3}x + 2$

Chapter 7
LOCUS

Locus (continued)

13. The locus of the center of a school bus wheel that is moving down a straight level road is:

    (1) one line

    (2) two lines

    (3) one circle

    (4) two circles

    _____

14. The perpendicular bisectors of the three sides of a triangle meet at a point that is:

    (1) equidistant from the three sides of the triangle

    (2) equidistant from the three vertices of the triangle

    (3) outside of the triangle

    (4) none of the above

    _____

15. The angle bisectors of a triangle meet at a point that is:

    (1) equidistant from the three sides of the triangle

    (2) equidistant from the three vertices of the triangle

    (3) outside of the triangle

    (4) do not meet at a point

    _____

16. Point R lies on $\overleftrightarrow{PQ}$. The locus of the center of all circles of radius 10 which pass through point R is:

    (1) one line

    (2) one circle

    (3) two parallel lines

    (4) two perpendicular lines

    _____

Chapter 7

LOCUS

Solving Problems Using Compound Loci

REMEMBER

To solve a compound locus problem sketch the required *loci* on the same diagram and count the number of points the loci intersect at.

Example: How many points are 3 centimeters from a given line and 4 centimeters from a point on the line?

Solution: (1) First draw the given line ℓ and place point P anywhere on the line.
(2) Sketch locus of points 3 cm from line ℓ.
(3) Sketch locus of points 4 cm from point P.
(4) Count the intersection of loci points.

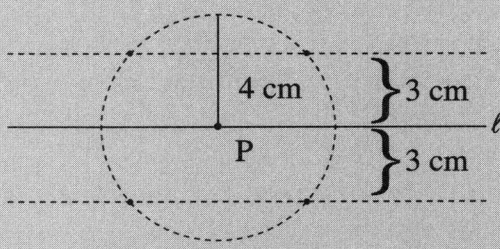

Solution: 4 points

1. What is the total number of points equidistant from two intersecting lines and 5 inches from their point of intersection?

    (1) 1   (3) 3

    (2) 2   (4) 4

2. What is the total number of points that are both 3 units from the y-axis and 3 units from the origin?

    (1) 1   (3) 0

    (2) 2   (4) 4

Chapter 7
LOCUS

Solving Problems Using Compound Loci (continued)

3. The distance between points P and Q is 6 units. How many points are equidistant from P and Q and also 3 units from P?

   (1) 1   (3) 3
   (2) 2   (4) 0

4. Points A and B are 8 units apart. What is the total number of points equidistant from A and B and also 4 units from the straight line passing through A and B?

   (1) 0   (3) 3
   (2) 2   (4) infinite number

5. What is the total number of points that are equidistant from two parallel lines and also equidistant from two points on one of the lines?

   (1) 1   (3) 3
   (2) 2   (4) 4

6. Points P and Q are 7 inches apart. How many points are 4 units from P and 3 units from Q?

   (1) 1   (3) 3
   (2) 2   (4) 0

7. Given: point P on line ℓ.
   - *a*   Describe fully the locus of points 4 inches from line ℓ.
   - *b*   Describe fully the locus of points d distance from point P.
   - *c*   Find the number of points that simultaneously satisfy the conditions in parts *a* and *b* above for the following values of d.

     1. d = 5
     2. d = 4
     3. d = 3

Chapter 7

LOCUS

Graph and Solve Compound Loci in the Coordinate Plane

REMEMBER

Example: Show graphically the numbered points 3 units from the coordinate (–3, 2) and 1 unit from the graph of y = 2. How many points satisfy both loci?

Solution: Graph $(x + 3)^2 + (y - 2)^2 = 4$ which is the locus of points 3 units from (–3, 2). On the same set of axes, graph y = 3 and y = 1 which is the locus of points 1 unit from y = 2.

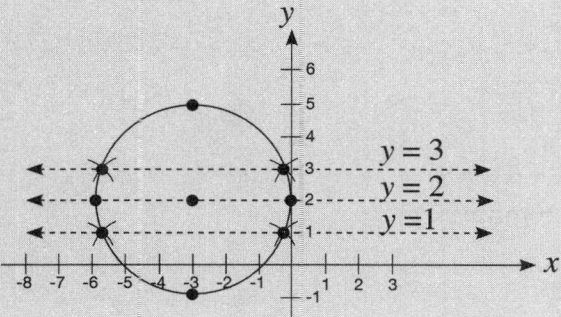

Answer: 4 points satisfy both loci.

1. Show graphically the number of points that are equidistant from the x and y axes and 3 units from the origin.

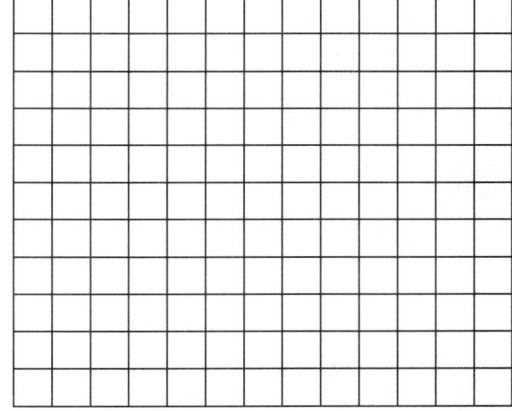

2. a How many points are equidistant from the lines y = 1 and y = –3 and 2 units from the point (2, 0)? (Show graphically.)
   b Find the coordinates of the point(s) in part a.

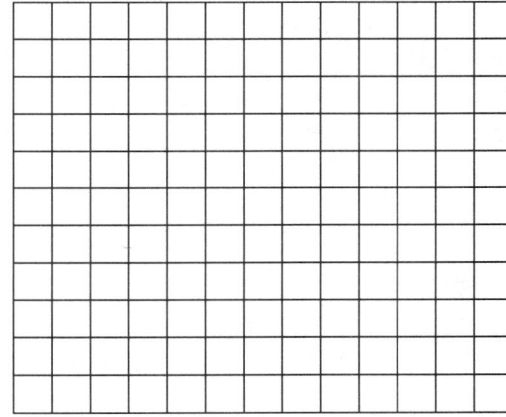

b_____

155

Chapter 7

LOCUS

Graph and Solve Compound Loci in the Coordinate Plane (continued)

3. What are the number of points that are equidistant from lines $x = -1$ and $x = 5$ and 3 units from the point (5, 2). (Show graphically)

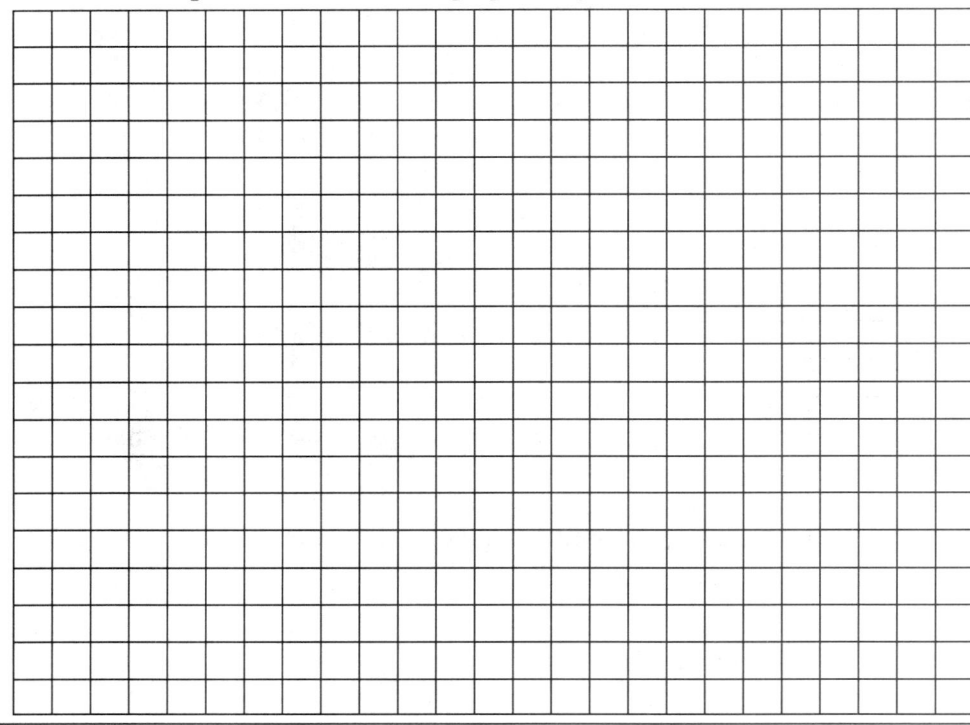

c _____

4. On the graph paper provided below:

   a Sketch the locus of points equidistant from two concentric circles with equations $x^2 + (y - 2)^2 = 4$ and $x^2 + (y - 2)^2 = 16$ consecutively.
   b On the same set of axes, sketch the locus of points that are 1 unit from the x axis.
   c How many points satisfy the loci in both parts *a* and *b*?

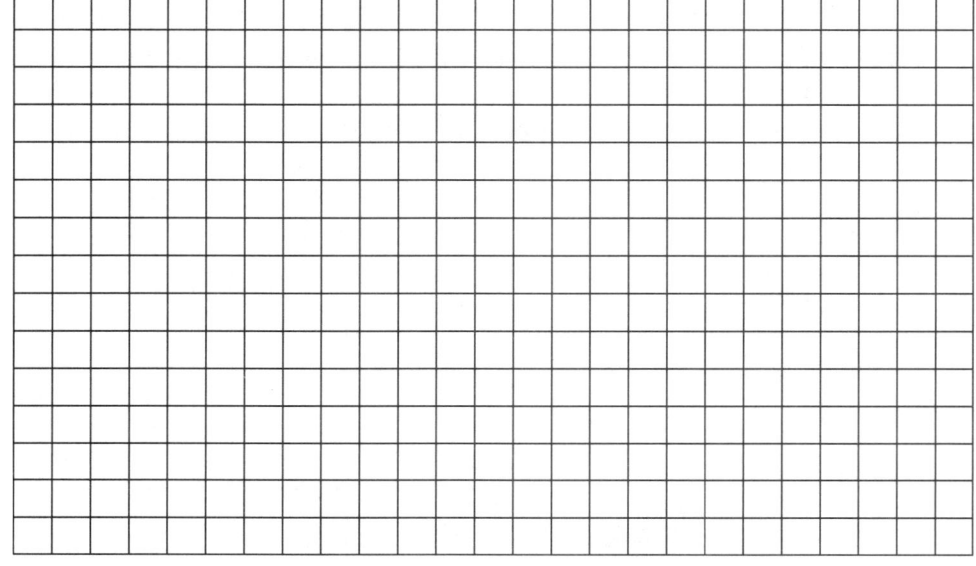

c _____

Chapter 7

LOCUS

Graph and Solve Compound Loci in the Coordinate Plane (continued)

5. a Draw the graphs of $x - 2y = 2$ and $2y - x = 8$ on the coordinate system provided below.

    b Determine the locus of points equidistant from the lines in part *a*. Draw this locus on the graph.

    c On the same set of axes, sketch the graph of the circle $(x - 2)^2 + (y - 2)^2 = 16$.

    d How many points of intersection do the graphs in part *b* and part *c* have in common?

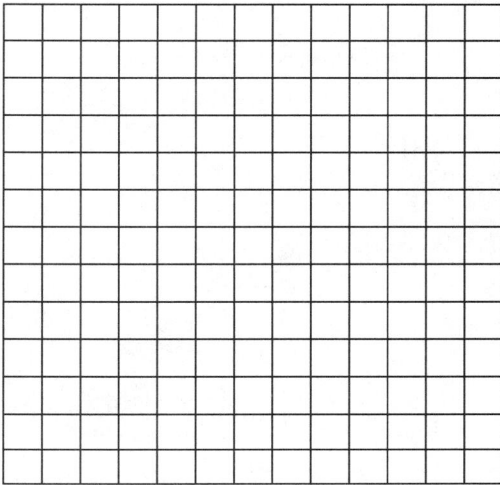

d_____

6. The equation of the locus of points 10 units from the point (–8 , 1) is

    (1) $(x - 8)^2 + (y + 1)^2 = 10^2$ (3) $(x + 8)^2 + (y - 1)^2 = 10^2$

    (2) $(x + 8)^2 + (y - 1)^2 = 10$ (4) $(x - 8)^2 + (y + 1)^2 = 10$

    _____

7. How many points are equidistant from the lines $y = 4$ and $y = -2$ and also equidistant from the points $A(-3 , 4)$ and $B(5, 4)$? Sketch the solution on the graph paper provided below.

_____

Chapter 8

CONSTRUCTIONS

Basic Constructions

> **REMEMBER**
>
> Review the procedures for basic geometric constructions. Depicted below are schematic diagrams for the most common constructions.
>
> *a*   copy an angle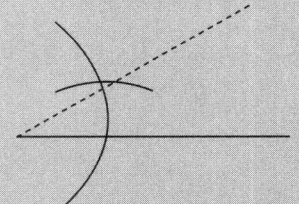
>
> *b*   bisect a line segment
>
> *c*   bisect an angle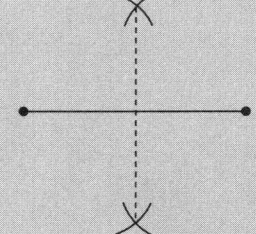
>
> *d*   drop a perpendicular from a point to a line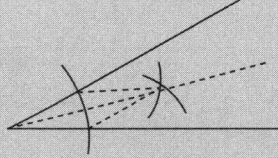
>
> *e*   erect a perpendicular at a point on a line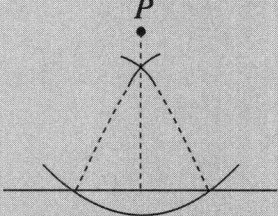
>
> *f*   construct a line parallel to a given line through a point outside of the given line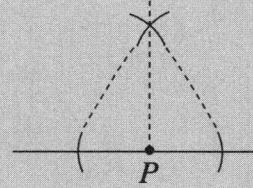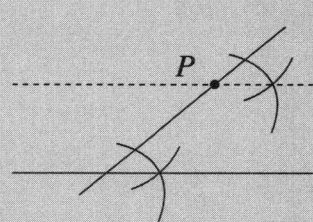

Chapter 8

CONSTRUCTIONS

Basic Constructions (continued)

1. Construct a line perpendicular to $\overleftrightarrow{ABC}$ through B. 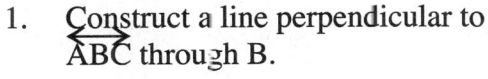	5. Construct a triangle whose sides are equal to length to a, b, and c. 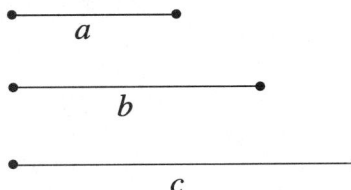
2. Construct and label $\overrightarrow{EG}$ such that $\angle$ FEG $\cong$ $\angle$ DEG. 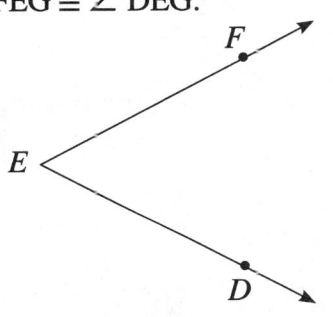	6. Construct the median to the non-parallel sides $\overline{AD}$ and $\overline{BC}$ of trapezoid ABCD 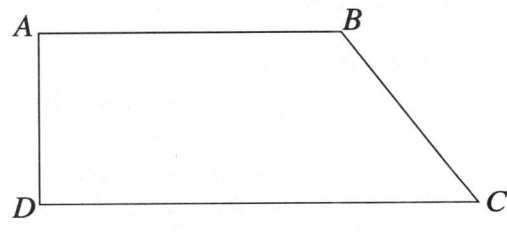
3. Through point A, construct a line parallel to $\overline{BC}$. 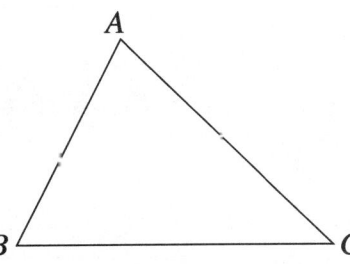	7. Construct a line perpendicular to $\overleftrightarrow{DEF}$ and passing through point P. 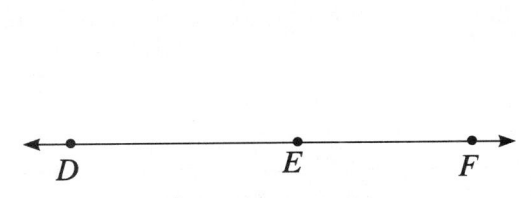 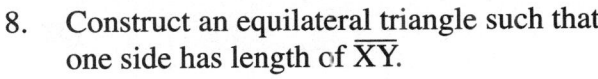
4. Construct the perpendicular bisector of the line segment connecting points P and Q.	8. Construct an equilateral triangle such that one side has length of $\overline{XY}$. 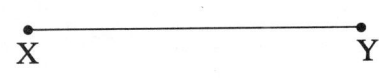

159

Chapter 8

CONSTRUCTIONS

Basic Constructions (continued)

9. In the accompanying figure, triangle ABC is scalene. The construction shows that $\overline{CD}$ is the

   (1) median to side $\overline{AB}$

   (2) bisector of angle C

   (3) altitude to side $\overline{AB}$

   (4) perpendicular bisector to side $\overline{AB}$

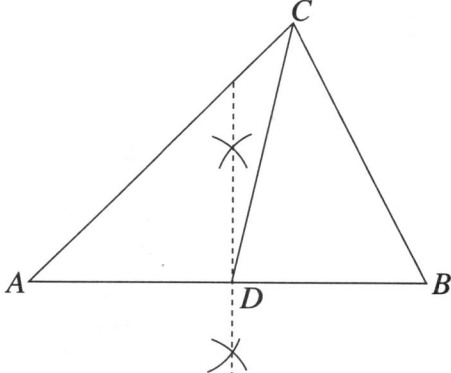

10. The accompanying diagram illustrates how to locate the center of a circle by construction methods. Which principle is used in the proof for this construction.

    (1) If two chords in the same circle are congruent, they are equidistant from the center of the circle.

    (2) The perpendicular bisector of a chord passes through the center of a circle.

    (3) An angle inscribed in a semicircle is a right angle.

    (4) A line through the center of a circle and perpendicular to a chord bisects the chord and its arcs.

11. Which drawing best illustrates the construction of an equilateral triangle?

(1)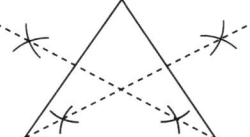

(3)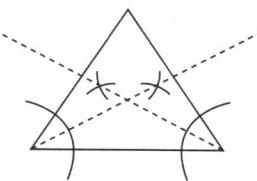

(2)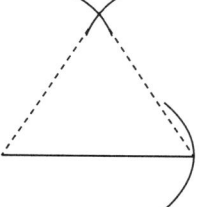

(4)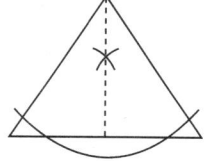

160

Chapter 8

CONSTRUCTIONS

Basic Constructions (continued)

12. In the accompanying diagram, △ ABC is scalene. According to the construction shown, what is $\overline{CD}$?

    (1) the altitude to side $\overline{AB}$
    (2) the bisector of angle C
    (3) the median to side $\overline{AB}$
    (4) the perpendicular bisector of side $\overline{AB}$

    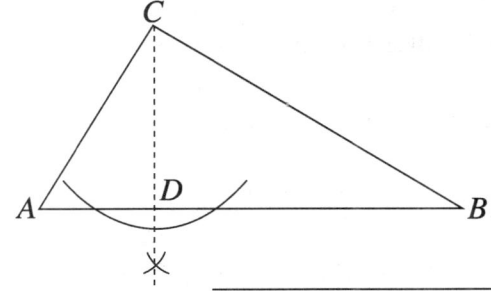

13. The diagram depicts the construction of line ℓ parallel to line m through point P. The reason line ℓ will be parallel to line m is

    (1) when two lines are cut by a transversal and a pair of congruent alternate interior angles are formed then the lines are parallel.
    (2) when two lines are cut by a transversal and a pair of congruent corresponding angles are formed then the lines are parallel.
    (3) parallel lines are everywhere equidistant.
    (4) a line perpendicular to one of two parallel lines will be perpendicular to the other also.

    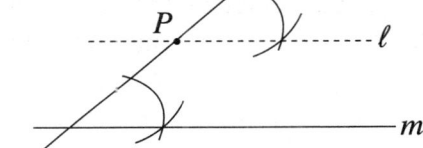

14. In constructing a perpendicular from an outside point P to line ℓ as shown below, the main reason that $\overline{PD}$ is perpendicular to line ℓ is that

    (1) two points equidistant from the endpoints of a line segment determine the perpendicular bisector of the segment
    (2) perpendicular lines form right angles
    (3) the shortest distance from a point to a line can never be determined
    (4) intersecting lines are either parallel or perpendicular.

    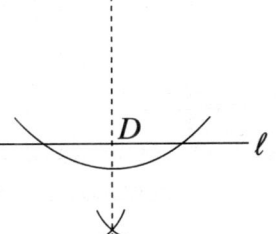

15. On a separate sheet of paper
    a   Draw three non-collinear points and label them A, B and C.
    b   Draw line segments $\overline{AB}$, $\overline{BC}$ and $\overline{AC}$.
    c   Construct the perpendicular bisectors of $\overline{AB}$ and $\overline{AC}$. Label the intersection of bisectors with the letter O.
    d   Draw a circle with the center at O and a radius of $\overline{OA}$.
    e   What conclusions can you arrive at concerning the circle O?

161

Chapter 8

CONSTRUCTIONS

Constructions (Circumcenter, Incenter and Centroid of a Triangle)

1. The circumcenter of a triangle is defined as the intersection of the perpendicular bisectors of the sides of the triangle. Using a straightedge and a compass, find by the basic constructions the circumcenter of triangle ABC described below.

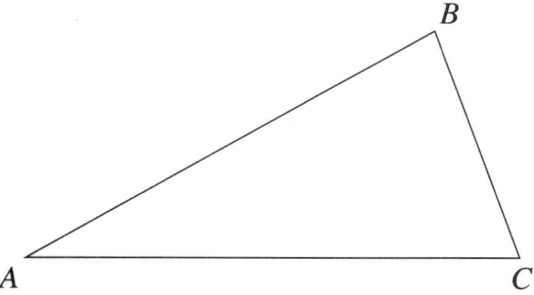

2. The incenter of a triangle is defined as the intersection of the angle bisector of the angles of the triangle. Using your construction implements, locate the incenter of triangle DEF.

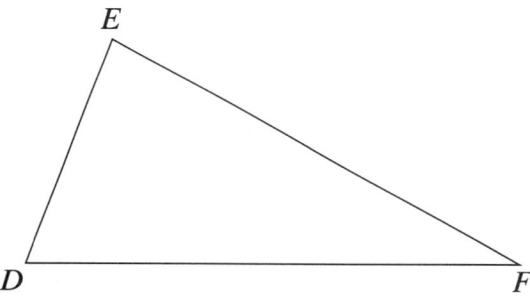

3. The centroid of a triangle is the intersection of the medians. With a straightedge and compass, locate the centroid of triangle JKL.

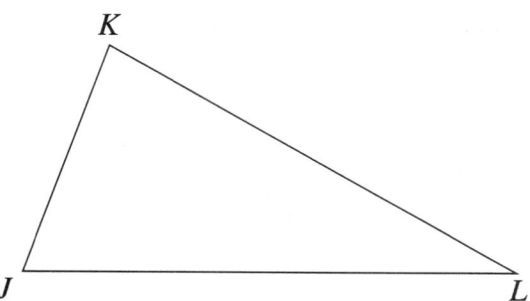

Chapter 8
CONSTRUCTIONS

Constructions (Orthocenter of a Triangle)

4. The orthocenter of a triangle is the intersections of the attitudes of the triangle. Using a straightedge and compass, locate the orthocenter of triangle RST.

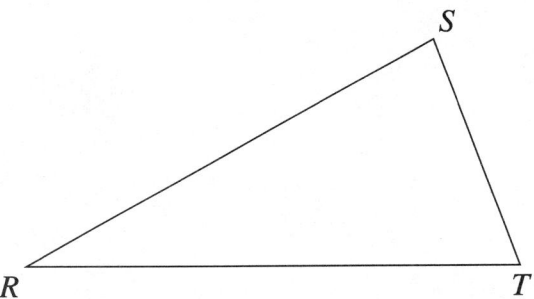

5. By construction, locate the orthocenter of triangle LMN.

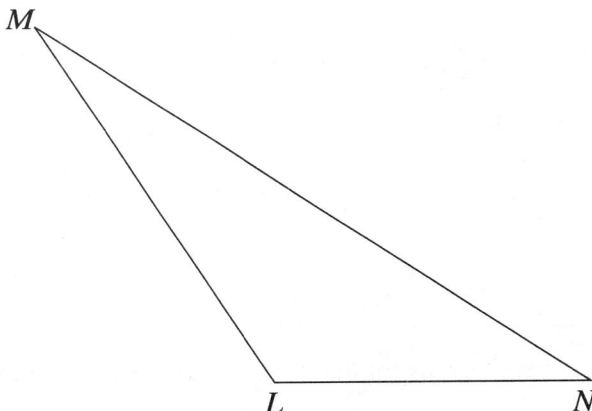

6. In which triangle would the circumcenter, incenter, centroid and orthocenter be the same point?

(1) an isosceles triangle

(2) an equilateral triangle

(3) an isosceles right triangle

(4) none of the above

Chapter 9

GEOMETRIC RELATIONSHIPS

A line perpendicular to each of two intersecting lines at their point of intersection

> **REMEMBER**
>
> A line perpendicular to each of two intersecting lines at their point of intersection, is perpendicular to the plane of the two intersecting lines.
>
>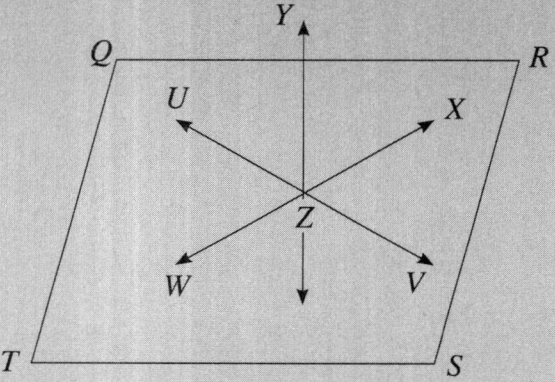
>
> For example, given $\overleftrightarrow{UV}$ intersecting $\overleftrightarrow{WX}$ at point Z,
> $\overleftrightarrow{YZ} \perp \overleftrightarrow{UV}$ and $\overleftrightarrow{YZ} \perp \overleftrightarrow{WX}$ and
> plane QRST passing through $\overleftrightarrow{UV}$ and $\overleftrightarrow{WX}$,
> it can be proven $\overleftrightarrow{YZ} \perp$ plane QRST.

1. The drawing below shows a pyramid whose base is a quadrilateral GEOM.

   Explain how you would prove that the line segment $\overline{TR}$ is the altitude of the pyramid.

   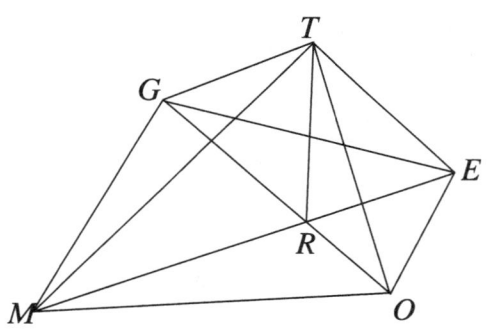

Chapter 9

GEOMETRIC RELATIONSHIPS

A plane perpendicular to a given line

> **REMEMBER**
>
> Through a given point in a given line, one plane, and only one plane, can be passed perpendicular to the given line.
>
>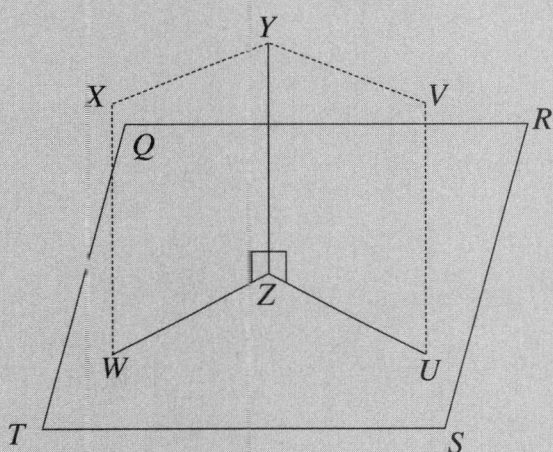
>
> For example, let $\overleftrightarrow{YZ}$ be the given line and Z the given point. Pass plane WXYZ through $\overleftrightarrow{YZ}$. In the plane WXYZ at the point Z, construct $\overleftrightarrow{WZ} \perp \overleftrightarrow{YZ}$. Pass a second plane UVYZ through $\overleftrightarrow{YZ}$ so that $\overleftrightarrow{ZU} \perp \overleftrightarrow{YZ}$. Pass plane QRST through the lines $\overleftrightarrow{WZ}$ and $\overleftrightarrow{ZU}$. No other plane can be passed through point Z and still be $\perp$ to line $\overleftrightarrow{YZ}$.

1. All the perpendiculars that can be drawn to a straight line at a given point, lie in the plane which is perpendicular to the given line at the given point.

    TRUE or FALSE?

    _____

2. Through a given external point, one plane, and only one plane, can be passed perpendicular to a given line.

    TRUE or FALSE?

    _____

165

Chapter 9

GEOMETRIC RELATIONSHIPS

A line perpendicular to a given plane

> **REMEMBER**
>
> Through a given point in a plane there passes one and only one line perpendicular to a given plane.
>
>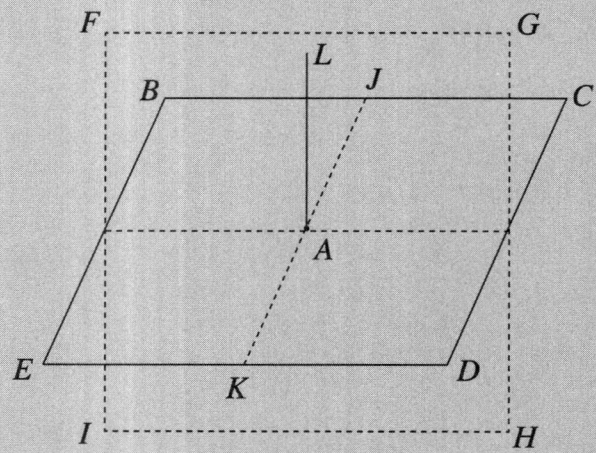

1. A carpenter is going to place the first vertical corner post for a deck on to a horizontal concrete floor. The horizontal concrete floor is represented by the rectangle ABCD and the vertical corner post is represented by line segment ED.

   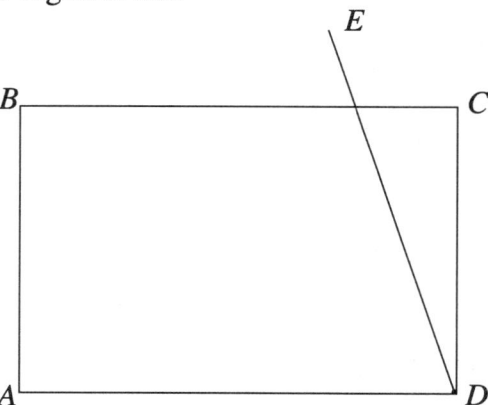

   Which angles in the figure must be made right angles in order for the post to be perpendicular to the floor?

   _____

Chapter 9

GEOMETRIC RELATIONSHIPS

Two lines perpendicular to the same plane

> **REMEMBER**
>
> Two lines perpendicular to the same plane are coplanar.
>
>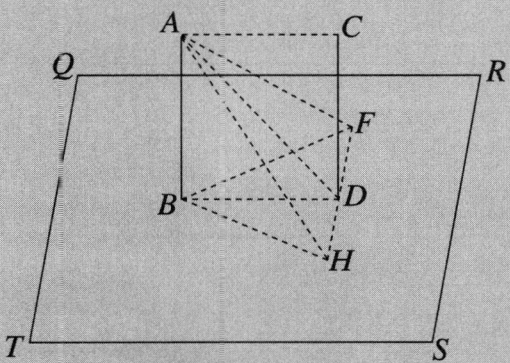
>
> For example, given $\overleftrightarrow{AB}$ and $\overleftrightarrow{CD} \perp$ plane QRST and meeting plane QRST at points B and D respectively, it can be proven $\overleftrightarrow{AB}$ and $\overleftrightarrow{CD}$ are coplanar.

1. Two lines perpendicular to the same plane are parallel.

   TRUE or FALSE ?

   _____

2. If one of two parallel lines is perpendicular to a plane, the other is also perpendicular to the plane.

       a SOMETIMES   *b* ALWAYS   *c* NEVER

   _____

3. If two lines are parallel to a third line, they are parallel to each other.

   TRUE or FALSE ?

   _____

4. Through a given external point there can be drawn one line perpendicular to a given plane, and only one.

   TRUE or FALSE ?

   _____

Chapter 9

GEOMETRIC RELATIONSHIPS

Two planes perpendicular to each other

REMEMBER

Two planes are perpendicular to each other if and only if one plane contains a line perpendicular to the second plane.

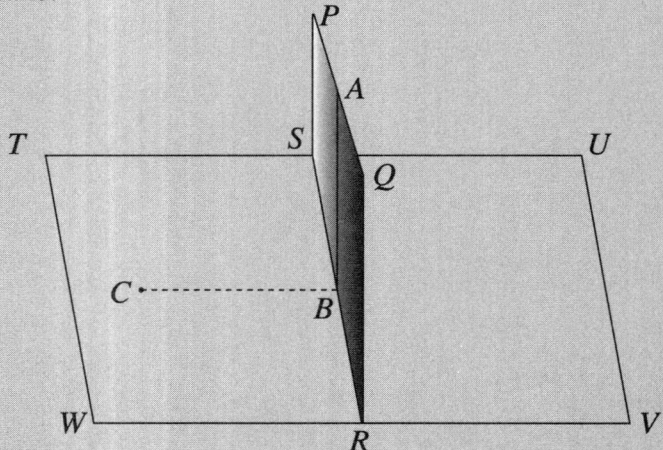

For example, If two planes, PQRS and TUVW. are perpendicular to each other, and a line $\overleftrightarrow{AB}$ is drawn in one of the planes, PQRS, and perpendicular to their intersection, line $\overleftrightarrow{SR}$,

it can be proven that line $\overleftrightarrow{AB}$ is perpendicular to the other plane, TUVW.

1. If two planes are perpendicular to each other, a line perpendicular to one of them at any point of their intersection will lie in the other plane.

   TRUE or FALSE ?

2. If two planes are perpendicular to each other, a line drawn perpendicular to one of them through any point in the other will lie in the second plane.

   TRUE or FALSE ?

3. If each of two intersecting planes is perpendicular to a third plane, their intersection, a line, is also perpendicular to that third plane.

       a SOMETIMES   *b* ALWAYS   *c* NEVER

Chapter 9

GEOMETRIC RELATIONSHIPS

A line perpendicular to a plane (Case #1)

REMEMBER

If a line is perpendicular to a plane, then any line perpendicular to the given line at its point of intersection with the given plane is in the given plane.

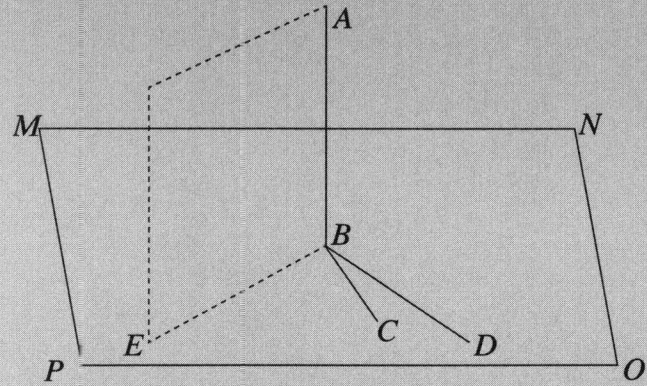

For example, given the plane MNOP and the line $\overleftrightarrow{BC}$, where both are ⊥ line $\overleftrightarrow{AB}$ at the point B,

It can be proven that line $\overleftrightarrow{BC}$ lies in the plane MNOP.

1. The perpendicular is the shortest line from a point to a plane.

    TRUE or FALSE ?

2. The distance from a point to a plane is the length of the perpendicular drawn from the point to the plane.

        a SOMETIMES   *b* ALWAYS   *c* NEVER

3. Explain why this fact is true:
    If one edge of a triangular prism is perpendicular to its base, then the prism is a right triangular prism.

169

Chapter 9

GEOMETRIC RELATIONSHIPS

A line perpendicular to a plane (Case #2)

REMEMBER

If a line is perpendicular to a plane, then every plane containing the line is perpendicular to the given plane.

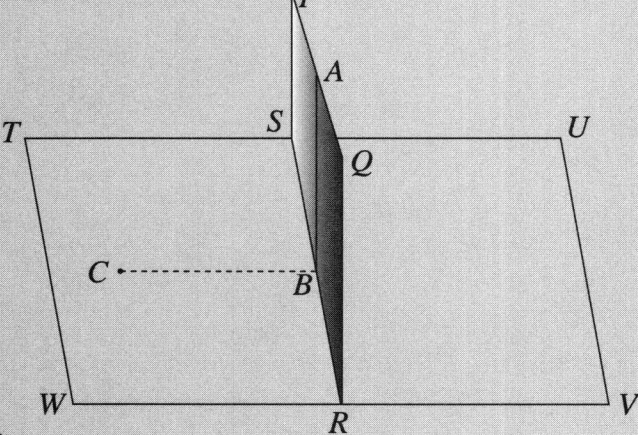

For example, if the line $\overleftrightarrow{AB} \perp$ plane TUVW, and the plane PQRS passes through line $\overleftrightarrow{AB}$ and intersects plane TUVW in segment $\overline{SR}$, it can be proven that plane PQRS $\perp$ plane TUVW.

1. A plane perpendicular to the edge of a dihedral angle is perpendicular to each of the two faces of the planes forming the dihedral angle.

   TRUE or FALSE ?

   Hint: A dihedral angle is the amount of opening between two intersecting planes.
   The faces of a dihedral angle are the planes forming the dihedral angle.
   The edge of a dihedral angle is the straight line in which the faces intersect.

   _____

2. a For each of the figures below, how many *vertical* symmetry planes exist?

   b For each of the figures below, how many *horizontal* symmetry planes exist?

   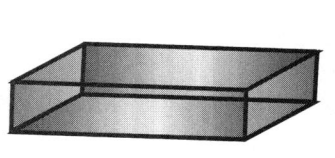      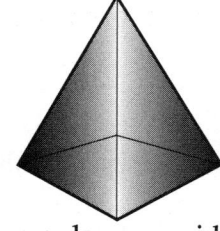   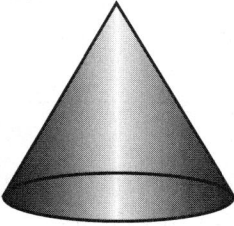

   rectangular solid   right prism   regular pyramid   right circular cone

   a _____   _____   _____   _____

   b _____   _____   _____   _____

170

Chapter 9

GEOMETRIC RELATIONSHIPS

A plane intersecting two parallel planes

> **REMEMBER**
>
> If a plane intersects two parallel planes, then the intersection is two parallel lines.
>
>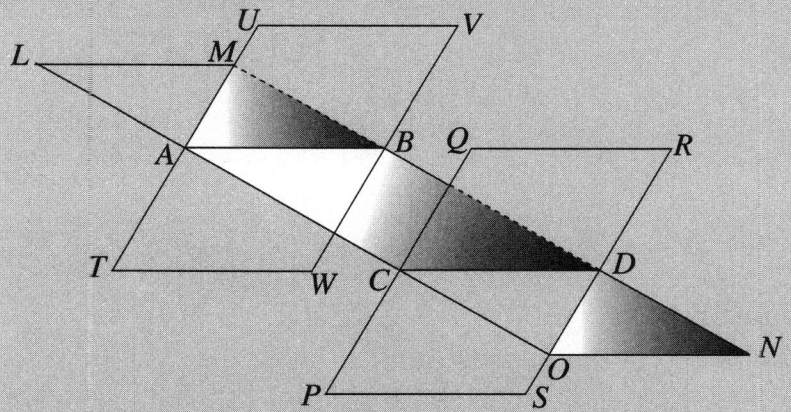
>
> For example, if planes TUVW and PQRS, two parallel planes, are intersected by the plane LMNO in the lines $\overleftrightarrow{AB}$ and $\overleftrightarrow{CD}$, respectively,
>
> It can be proven that line $\overleftrightarrow{AB}$ is parallel to line $\overleftrightarrow{CD}$.

1. Parallel lines included between parallel planes are congruent.

    a SOMETIMES   *b* ALWAYS   *c* NEVER

    _____

2. A line perpendicular to one of two parallel planes is perpendicular to the other plane also.

    TRUE or FALSE ?

    _____

3. Through a given point outside a plane, one plane, and only one, can be passed parallel to the given plane.

    TRUE or FALSE ?

    _____

4. Two parallel planes are everywhere equidistant.

    a SOMETIMES   *b* ALWAYS   *c* NEVER

    _____

Chapter 9

GEOMETRIC RELATIONSHIPS

Two planes perpendicular to the same line

> **REMEMBER**
>
> If two planes are perpendicular to the same line, they are parallel.
>
>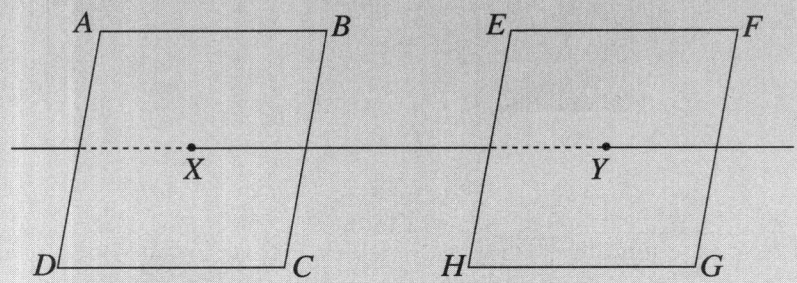
>
> For example, if the planes ABCD and EFGH are perpendicular to line XY, it can be proven that plane ABCD is parallel to plane EFGH.

1. The figure below shows two right hexagonal prisms.

   a. On one figure, sketch a symmetry plane and write a description of the symmetry plane that uses the word *parallel*.

   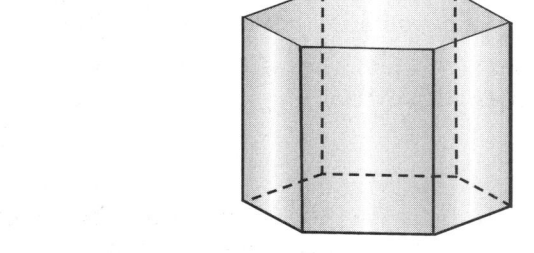

   b. On one figure, sketch a symmetry plane and write a description of the symmetry plane that uses the word *perpendicular*.

   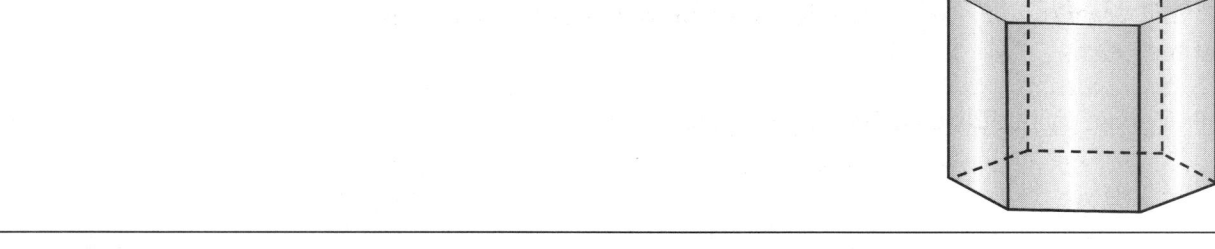

2. Given the line AB as shown below. Use two pieces of paper to construct two planes that are squares with sides of 2 inches. Hold these two planes such that they are perpendicular to line AB.

   Make a conjecture regarding those two planes and explain your conjecture.

172

Chapter 9

GEOMETRIC RELATIONSHIPS

Lateral edges of a prism

> **REMEMBER**
> The lateral edges of a prism are congruent and parallel. They are parallel lines included between parallel planes and are also congruent since parallel lines included between parallel planes are congruent.

1. A *polyhedron* is a solid bounded by planes. TRUE or FALSE?	_____
2. The *faces of a polyhedron* are its bounding planes. TRUE or FALSE?	_____
3. The *edges of a polyhedron* are the lines of intersection of the faces. TRUE or FALSE?	_____
4. A *prism* is a polyhedron bounded by two parallel planes and a group of planes whose lines of intersection are parallel. *a* SOMETIMES *b* ALWAYS *c* NEVER	_____
5. The *bases of a prism* are the faces formed by the two parallel planes bounding the prism. TRUE or FALSE?	_____
6. The *lateral faces* are the faces formed by the group of planes bounding the prism whose lines of intersection are parallel. *a* SOMETIMES *b* ALWAYS *c* NEVER	_____
7. The *altitude of a prism* is the perpendicular distance between the planes of the bases of the prism. TRUE or FALSE?	_____
8. The *lateral faces* of a prism are parallelograms. TRUE or FALSE?	_____
9. The *bases of a prism* are congruent polygons. *a* SOMETIMES *b* ALWAYS *c* NEVER	_____
10. A *right prism* is a prism whose lateral edges are perpendicular to the base. TRUE or FALSE?	_____

Chapter 9

GEOMETRIC RELATIONSHIPS

Two prisms with equal volumes

> **REMEMBER**
>
> Two prisms have equal volumes if their bases have equal areas and their altitudes are equal.
>
> **Example 1:** A prism with a triangular base has a base area of x square units and an altitude of y units. A second prism wit a regular hexagonal base has a base area of x square units and an altitude of y units. What is the ratio of the volume of the first prism to the volume of the second prism?
>
> **Solutions 1:** Two prisms have equal volumes if their bases have equal areas and their altitudes are equal. Therefore the ratio of volumes is 1:1. Answer
>
> **Example 2:** A prism with a triangular base has an altitude = 25.0 cm. The base area = 18.0 cm². A second prism with a regular hexagonal base has an altitude of 25.0 cm and a volume equal to the first prism. Find the side of the regular hexagonal base to the *nearest hundredth* of a cm.
>
> **Solution 2:** $V_1 = Bh = 18.0(25.0) = 450.0 \text{ cm}^3$
>
> $V_1 = V_2$; $450.0 = B(25.0)$; $B = 18.0 \text{ cm}^2$
>
> $K = \dfrac{1}{2}ap$ where K = area, a = apothem, p = perimeter
>
> $18.0 = \dfrac{1}{2}(\dfrac{1}{2}s\sqrt{3})(6s)$; in a regular hexagon, the apothem $= \dfrac{1}{2}s\sqrt{3}$
>
> $s^2 = \dfrac{18.0(4)\sqrt{3}}{6(3)}$; $s = 2.63$ cm Answer

1. A rectangular prism has an altitude of 15 inches and a base area of 31.2 sq in. A second rectangular prism has a square base, an altitude of 15 in., and the same volume as the first prism. Find the side of the square to the *nearest tenth* of an inch.

Chapter 9

GEOMETRIC RELATIONSHIPS

Two prisms with equal volumes (continued)

2. A rectangular prism has an altitude of 4 inches and one side of 2 inches. A cube has an edge with a measure of 4 inches. The rectangular prism and the cube have equal volumes. Find the value of the third edge of the rectangular prism.

3. A solid wooden pedestal has the form of a regular hexagonal prism whose base edge is 8 *inches* and whose height is 15 *feet*. Find, to the nearest *pound*, the weight of the pedestal if a rectangular block of the same material, 2 *inches* by 10 *inches* by 3 *feet*, weighs 20 pounds.

4. A rectangular solid has a base area of 43.3 sq inches and a height of 50 inches. A triangular prism has a base in the shape of an equilateral triangle and a height of 50 inches. The volumes of the rectangular solid and triangular prism are equal. Find the side of the equilateral triangle for the base of the prism to the nearest inch.

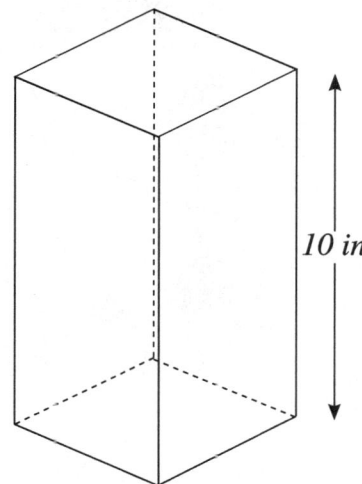

10 in

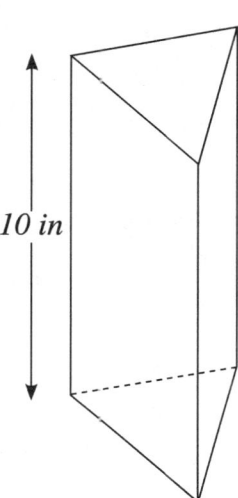

10 in

Chapter 9

GEOMETRIC RELATIONSHIPS

Volume of a prism

> **REMEMBER**
>
> The volume of a prism is the product of the area of the base and the altitude.
>
> VOLUME = BASE AREA × ALTITUDE
> V = Bh   (where B is base area and h is altitude)
>
> The lateral area of a right prism is the product of the perimeter of the base and the altitude.
>
> LATERAL AREA = BASE PERIMETER × ALTITUDE
> S = ph   (where S is lateral area, p is perimeter and h is altitude)
>
>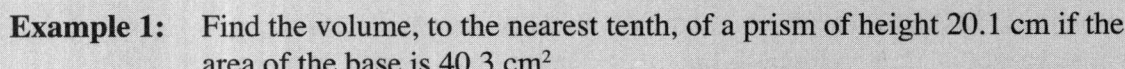
>
> **Example 1:** Find the volume, to the nearest tenth, of a prism of height 20.1 cm if the area of the base is 40.3 cm².
>
> **Solution 1:** V = Bh = 40.3 × 20.1 = 810.0 cm³   Answer
>
> **Example 2:** A right prism has a square base with a side of length = 3.8 inches. The volume of the right prism is 673.8 in³. Find the height to the nearest tenth.
>
> **Solution 2:** V = Bh
> 673.8 = 3.8 × 3.8 × h
> h = 46.7 inches   Answer
>
> **Example 3:** Find the lateral area and volume of a rectangular solid having a length of 2.7 yd, width of 9.1 yd and a height of 12.3 yd. Round answers to nearest tenth.
>
> **Solution 3:** S = ph = 2(2.7)(9.1) × 12.3 = 604.4 yd²   Answer
> V = Bh   (2.7)(9.1) × 12.3 = 302.2 yd³   Answer
>
> **Example 4:** Find the volume of a cube whose lateral area is 144 square units.
>
> **Solution 4:** S = ph = 4e × e = 4e² square units [where e is the edge or side of the cube]
> 144 = 4e²,   e² = 36,   e = 6.0 units   [where e is the cube edge or side]
> V = Bh = 36(6) = 216 cubic units   Answer

Chapter 9

GEOMETRIC RELATIONSHIPS

Volume of a prism (continued)

1. The base of a right prism is a rhombus whose diagonals are 12 cm and 16 cm. The altitude of the prism is 20 cm. Find its lateral area.

2. Find the volume of a regular hexagonal prism whose base edge is 6 inches and whose altitude is 10 inches. Round answer to nearest cubic inch.

3. If two storage bins have shapes of rectangular solids, find the larger volume?
   A storage bin 10ft. by 2.4 ft. by 7 ft. or a storage bin 8.1 ft. by 4.2 ft. by 5.3 ft.?

4. Find the number of cubic centimeters in the volume of a cube whose edge is 34.7 cm.

5. Find the volume of a cube whose lateral area is 676 square units.

6. If the altitude of a triangular prism is 81 in, and the right triangular base has legs of 10 in. and 24 in., find the volume.

7. Find the volume of a triangular prism whose altitude is 48 cm and the edges of its triangular base have lengths of 14 cm, 16 cm and 18 cm.

8. If the volume of a triangular prism is 206.64 cu ft and the height is 10.33 ft, find the base area.

9. Find the volume of a prism whose base area is 65.43 sq cm and altitude is $34\frac{1}{3}$ cm.

10. Find the volume of a prism whose base is an isosceles right triangle with a leg = 7 in. and altitude = 10 inches.

Chapter 9

GEOMETRIC RELATIONSHIPS

Volume of a prism (continued)

11. How many 12 *cu yd* dump trucks would be required to remove the earth from a building site where a diving pool 40 ft by 30 ft by 15 ft is to be made.

12. The perimeter of a right section of a prism is 16 cm. and the lateral edge is 20 cm. Find the lateral area of the prism.

13. The altitude of a prism is 9 units and its base is a right triangle whose legs are 3 and 4 units. Find the prism volume.

14. If the 6 sides of a cube have a total area of 150 sq yd, find its volume.

15. The altitude of a prism is 8 meters and the base is an equilateral triangle with a side of 6 meters. Find the volume, of the prism to the nearest integer.

16. A regular triangular prism has a height of 24 in. and a base with one side = 5 in. Find its volume to the nearest cubic inch.

17. Find the volume of a prism if a right section of the prism is a rectangle 3 ft by 8 ft and its lateral edge is 10 ft.

Chapter 9

GEOMETRIC RELATIONSHIPS

Volume of a prism (continued)

18. The figure at the right represents a regular hexagonal spacer. The radius of the hole is 0.5 inch. The edge of the spacer is 1 inch and its thickness is 0.5 inch. Find, to the nearest tenth of a cubic inch, the volume of the spacer.

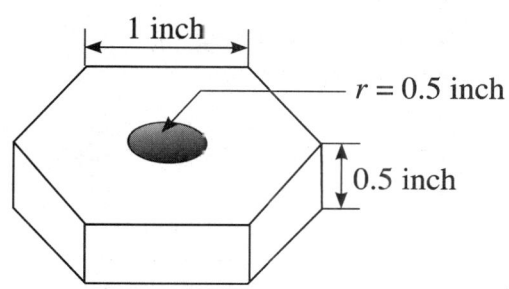

19. A regular hexagonal prism and a right circular cylinder each have altitudes of 20 feet. An edge of the base of the prism is 2 feet and the radius of the cylinder is 1 foot. Find, to the nearest cubic foot, the difference between their volumes.

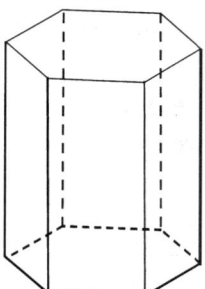 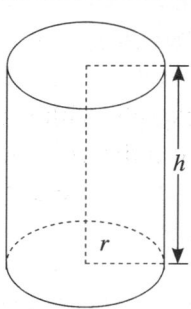

Chapter 9

GEOMETRIC RELATIONSHIPS

Volume of a prism (continued)

20. A cross section of a railroad embankment is in the shape of an isosceles trapezoid whose bases are 20 feet and 36 feet. The congruent legs are 17 feet each. If the railroad embankment is 300 feet long, how many *cubic yards* of earth have been used?

21. A storage building has the form of a right prism with a base in the shape of an isosceles triangle. The dimensions of two sides of the base are 40 ft. and 90 ft. The altitude of the right prism is 250 ft. Find the volume of the right prism.

Chapter 9

GEOMETRIC RELATIONSHIPS

Volume of a prism (continued)

22. A solid column of wood has the form of a regular hexagonal prism whose base edge is 1.0 ft. and whose altitude is 25.0 ft. Find the volume of the solid column.

23. A deck for a train station requires 1200 concrete posts. The posts are to have the form of a prism whose base is a square 8 inches on a side and whose height is 6 ft. The concrete has a ratio of 1:2:4 for cement, sand and gravel respectively. Find, to the nearest cubic foot, the amount of cement needed to make 1200 concrete posts.

Chapter 9

GEOMETRIC RELATIONSHIPS

Volume of a prism (continued)

24. A recycling plant takes bars of metal each having a form of a regular square prism with dimensions 5 in. by 5 in. by 84 in. and melts them. It then takes the melted product and casts it into cylinder solids each having the form of a hollow cylinder. The hollow cylinder has a length of 2 in., an outer radius of 1 in. and an inner radius of 0.75 in. How many hollow cylinders can be made from one bar if there is no waste?

25. A swimming pool in the shape of a rectangle prism has dimensions 33 feet long and 21 feet wide. The pool has two depths. The swimming area depth is 3 feet and the diving area depth is 12 feet. The length for diving to length for swimming is 1:3 respectively. Find the swimming and diving surface area.

   a Find the volume of water required in the swimming section of the pool.
   b Find the volume of water required in the diving section of the pool.

26. Two cereal boxes are in the shape of a rectangular prism. Box "A" is 8.00 in. by 5.75 in. by 2.75 in. Box "B" is 7.25 in. by 4.75 in. by 3.00 in.

   a Find the volume of both boxes of cereal to the nearest hundredth of a cubic inch.
   b Which box is the better buy if they sell at the same price?
   c If Box "A" sells for "y" dollars, what would the equivalent price be, in terms of "y" dollars for Box "B"?

Chapter 9

GEOMETRIC RELATIONSHIPS

A regular pyramid

> **REMEMBER**
>
> Properties of a regular pyramid:
>   Lateral edges are congruent.
>   Lateral faces are congruent isosceles triangles.
>   Volume of a pyramid equals
>   one-third the product of the
>   area of the base and the altitude.
>
> $$\text{VOLUME} = \frac{1}{3}(\text{BASE AREA} \times \text{ALTITUDE})$$
>
> $$V = \frac{1}{3}Bh$$
>
> $$\text{LATERAL AREA} = \frac{1}{2}(\text{BASE PERIMETER}) \times \text{SLANT HEIGHT}$$
>
> $$S = \frac{1}{2}pl$$
>
> **Example 1:** The volume of a pyramid with a square base is 108 cu inches. If the measure of the altitude is 4 inches, what is the length of a side of the base?
>
> **Solution 1:**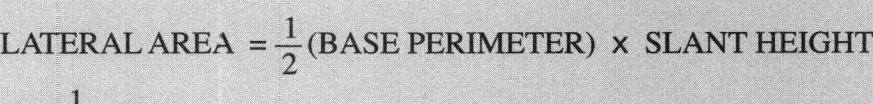
>
> $108 = \frac{1}{3}(s)^2(4)$ $\qquad$ $s = 9$ inches $\qquad$ Answer
>
> **Example 2:** A regular pyramid has a square base, one side of which has a measure of 12 cm. The altitude of the pyramid has a measure of 8 cm. Find the length of the altitude of a triangular face known as the slant height, l.
>
> **Solution 2:** The slant height, l, is the hypotenuse of a right triangle with the altitude, h, as one leg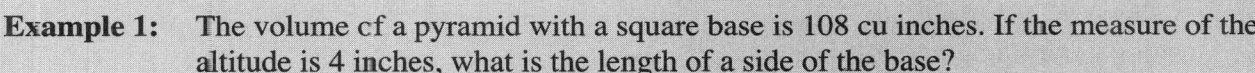
> $a^2 + b^2 = c^2$
> $6^2 + 8^2 = l^2$
> slant height, $l = 10$ cm $\qquad$ Answer
>
> **Example 3:** Find the lateral area of the regular pyramid in example 2 above using the slant height found.
>
> **Solution 3:** $S = \frac{1}{2}pl = \frac{1}{2}(4)(12)(10) = 240$ cm² $\qquad$ Answer
>
> **Example 4:** Express the volume of a regular square pyramid in terms of its altitude, h, and its base edge, e.
>
> **Solutions 4:** $V = \frac{1}{3}Bh = \frac{1}{3}e^2h$ $\qquad$ Answer

183

Chapter 9

GEOMETRIC RELATIONSHIPS

A regular pyramid (continued)

1. Find the volume of a pyramid if the base area = 48.23 sq in and altitude = 39.7 in.

   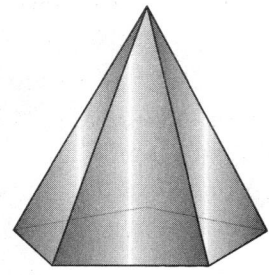

2. If the volume of a pyramid = 546 cm³ and the altitude = 97.7 cm, find the area of the base.

3. Find the volume of a pyramid with a base that is an equilateral triangle with a side of 7.2 ft and an altitude of 21 ft.

4. Find the volume of a pyramid with a base that is a right triangle with hypotenuse = 56 in and one leg = 27 in. The altitude of the pyramid is 72 in.

5. If a pyramid has an altitude of 263 yd and a square base with a side of 312 yd, how many cubic yards of material does it consist of?

Chapter 9

GEOMETRIC RELATIONSHIPS

A regular pyramid (continued)

6. A cell tower is a hexagonal pyramid 250 ft high. Each side of the base is 12 ft. This cell tower has a hollow interior in the shape of a hexagonal pyramid with each side of the base = 10 ft. The hollow interior has a height of 85 ft. Find the volume, to the nearest cu ft, of the material making up the cell tower.

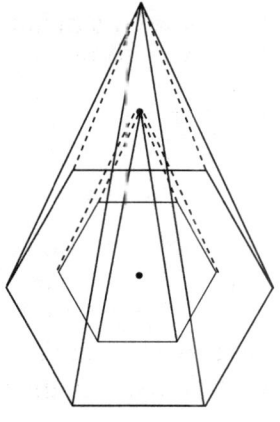

7. If a pyramid has a volume of 124 *cu yd* and its base is a square with a side of 17 *ft*, find the altitude to the nearest foot.

   [Hint: 27 cu ft = 1 cu yd]

8. Find the volume of a regular square pyramid whose slant height = 72 units and side of the base = 46 units.

9. Find the lateral area of a regular pyramid with a regular hexagonal base of sides each = 68 cm and slant height = 489 cm.

185

Chapter 9

GEOMETRIC RELATIONSHIPS

A regular pyramid (continued)

10. If a pyramid is cut by a plane parallel to its base, the section formed is similar to the base.
       a ALWAYS   *b* SOMETIMES   *c* NEVER

11. The base edges of a frustum of a regular square pyramid are 3 units and 7 units and the slant height is 6 units. Find the lateral area of the frustum.

12. The altitude of a pyramid is 20 inches. Its base is a square each of whose sides is 5 inches long. Find the area of the section made by a plane parallel to the base and 8 inches from the vertex.

13. Find the lateral area of a regular pentagonal pyramid whose base edge is 4 units and whose slant height is 10 units.

14. A plane is parallel to the base of a pyramid and bisects the altitude. Find the ratio of the area of the section formed to the area of the base.

Chapter 9

GEOMETRIC RELATIONSHIPS

A regular pyramid (continued)

15. Express the volume of a regular square pyramid in terms of its altitude, h and its base edge, e.

16. The altitude of a pyramid is 6 inches and the base is a right isosceles triangle with legs of 6 inches. Find the volume of the pyramid.

17. The area of the base of a pyramid is 30 sq ft and its volume is 120 cu ft. find its altitude.

18. A pyramid with a square base 15 feet on a side is 60 feet high. Find the area of the section parallel to the base and 20 feet from the base.

19. The total area of a regular tetrahedron is $36\sqrt{3}$. Find an edge of the tetrahedron.
    HINT: The *regular* tetrahedron has four congruent equilateral triangles.

Chapter 9

GEOMETRIC RELATIONSHIPS

A regular pyramid (continued)

20. The lateral faces of a square pyramid make with the base an angle of 45°. If the altitude of the pyramid is h, express the volume of the pyramid in terms of h.

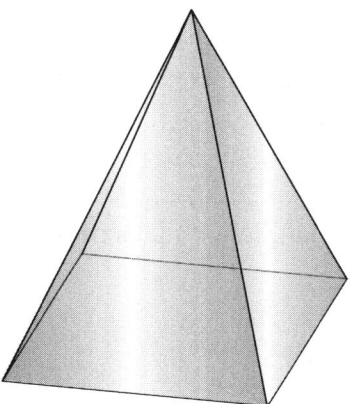

21. A swimming pool hopper is required for diving. The hopper consists of two parts as shown in the drawing. The upper part has the form of a rectangular solid whose height is 5 feet. The lower part has the form of a pyramid whose height is 4 feet. The common base is 16 feet long and 9 feet wide. Find, to the nearest cu ft, the capacity of the hopper.

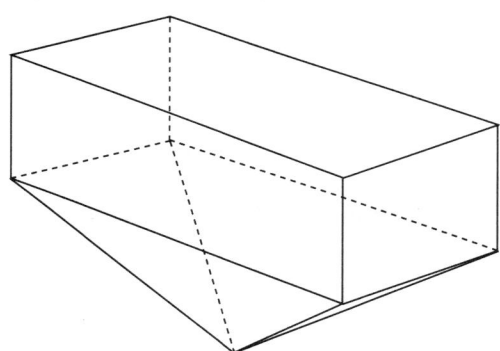

Chapter 9

GEOMETRIC RELATIONSHIPS

A regular pyramid (continued)

22. The base of a pyramid is isosceles trapezoid ABCD with $\overline{AB} \parallel \overline{DC}$, AB = 17 units, CD = 33 units, and AD = 10 units.
    Find the length of the altitude of trapezoid ABCD.
    Find the area of trapezoid ABCD.
    Find the distance, x from the vertex of the pyramid to a cross section of it if the area of the cross section is 54 sq units and the length of the altitude of the pyramid is 15 units.

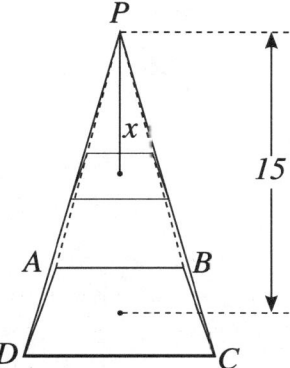

23. The lateral area of the regular square pyramid P-ABCD is 260 sq units, and the perimeter of the base is 40 units.
    Find the measure of the slant height $\overline{PE}$.
    Find the measure of the altitude $\overline{PO}$.
    Find the volume of the pyramid.
    Find the measure of the lateral edge $\overline{PA}$.
    Find the volume of the pyramid cut off by a plane parallel to the base at a distance of 6 units from the vertex.

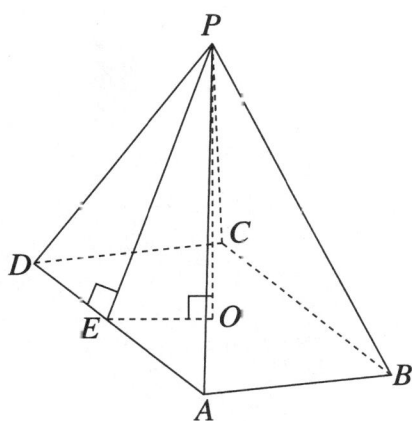

Chapter 9

GEOMETRIC RELATIONSHIPS

A Cylinder

> **REMEMBER**
>
> Properties of a cylinder:
>
> Bases are congruent.
>
> Volume of a cylinder equals the product of the area of the base and the altitude.
>
> VOLUME = BASE AREA x ALTITUDE
>
> $V = Bh$
>
>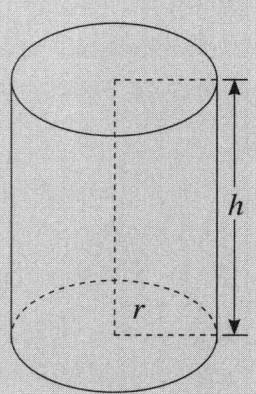
>
> Lateral area of a right circular cylinder equals the product of the circumference of the base and the altitude
>
> LATERAL AREA = CIRCUMFERENCE of BASE x ALTITUDE
>
> $S = 2\pi rh$
>
> **Example 1:** Find the volume of a cylinder with a base area of 24.9 sq in and an altitude of 10 in.
>
> **Solution 1:** $V = Bh = 24.9 (10) = 249$ cu in   Answer
>
> **Example 2:** Find the lateral area of a right circular cylinder with a radius of 50 cm and a height of 32.8 cm. Round the answer to nearest tenth.
>
> **Solution 2:** $S = 2\pi rh = 2\pi (50)(32.8) = 10304.4$ sq cm
>
> **Example 3:** If the volume of a cylinder is 500 ft³, find the base area of the cylinder if the base area to height ratio is 5:4
>
> **Solution 3:** Let $B = 5x$ and $h = 4x$
> $V = Bh$ $500 = (5x)(4x)$ $20x^2 = 500$
> $x = 5$
> Base area = $B = 5x = 25$ ft² and height = $h = 4x = 20$ ft   Answer

Chapter 9

GEOMETRIC RELATIONSHIPS

A Cylinder (continued)

1. The radius of the base of a right circular cylinder is 4 ft and the altitude of the cylinder is 6 ft. Find the volume of the cylinder.

2. The lateral surface area of a right circular cylinder is 60π cm². If the height of the cylinder is 10 cm, find the radius of the cylinder.

3. A cylinder and a cone have equal volumes and equal altitudes. If the area of the base of the cylinder is 18 sq in, find the area of the base of the cone.

4. Express in terms of r the total area of a right circular cylinder whose radius is r and whose altitude is $3r$.

5. The altitude of a cylinder of revolution is twice the radius of the base. Find the ratio of its lateral area to its total area.

6. The radius of the base of a right circular cylinder is 6 ft and its altitude is 14 ft. Find the total area in terms of π.

7. The lateral area of a cylinder of revolution is 240π sq in and its altitude is 10 inches. find the radius of the base.

8. If the radius of a right circular cylinder is doubled without changing its altitude, by what number is its volume multiplied?

Chapter 9

GEOMETRIC RELATIONSHIPS

A Cylinder (continued)

9. Express the total area of a right circular cylinder as a function of its radius r and its altitude h.

10. Find the lateral area of a right circular cylinder whose radius is 4 units and whose altitude is 9 units.

11. The radius of a circular cylinder is r and its altitude is $3r$. Express the volume of the cylinder in terms of r.

12. Express the total area of a right circular cylinder of radius r and altitude $2r$.

13. In the formula $V = Bh$, let $V = 72.75$ and $r = 3.25$. Find h to the nearest hundredth.

14. Two similar cylindrical tanks have depths of 12 ft and 6 ft. What is the ratio of the volume of the larger tank to the volume of the smaller tank?

15. Find the volume, to the nearest integer, of the cylinder having a diameter of 30 ft and a height of $7\frac{2}{5}$ yd.

16. How many 9.5 cu ft barrels are required when emptying a cylindrical tank 22.5 ft long and 9 ft in diameter?

17. A tunnel has a cross sectional area of a semicircle with a radius of 10 yd. The tunnel is 1760 yd long. Find the volume of the tunnel to the nearest cu yd.

Chapter 9

GEOMETRIC RELATIONSHIPS

A Cylinder (continued)

18. If a *cubic foot* of copper is made into a solid copper wire with a diameter of 0.1019 *inch*, how long, to the *nearest foot*, will the wire be?

19. A coffee can is in the shape of a cylinder with a diameter of 4 in and a height of 5.5 in. It holds 11.5 oz of coffee when full. A second coffee can, cylindrical in shape, has a diameter of 6.125 in and a height of 6.25 in. Find, to the *nearest tenth of an ounce*, the weight of coffee the larger can if able to hold when it is full.

20. A cooking spray can is in the shape of a cylinder with a diameter of 2.25 in and a height of 6 in. The can holds 5 oz. The manufacturer would like to package a 12 oz cylindrical can of cooking spray with a diameter of 2.5 in. What would the height of the 12 oz can be to the nearest tenth of an inch?

Chapter 9

GEOMETRIC RELATIONSHIPS

A Cylinder (continued)

21. The volume of a sphere is equal to that of a right circular cylinder whose diameter d is equal to its altitude. Express radius r of the sphere as a function of d.

   Find r to the nearest tenth if $d = 12$.

22. Before sharpening, a pencil was in the form of a right circular cylinder 7 inches long and $\frac{1}{4}$ inches in diameter. Sharpening it to a conical point reduced its overall length to $6\frac{7}{8}$ inches with the cylindrical part becoming 6 inches long. Find, to the nearest hundredth of a cubic inch, the volume of the material removed.

23. Find to the nearest pound, the weight of 4 feet of pipe which has a 2 inch inside diameter and is $\frac{1}{4}$ inch thick. The pipe material weighs 708 pounds per cu ft.

Chapter 9

GEOMETRIC RELATIONSHIPS

A Cylinder (continued)

24. How many feet of plastic piping can be made from 5170 cu in of plastic if the outer diameter of the pipe is 8 inches and the thickness of the pipe is 1 inch?

25. Fruit juice is sold in cylindrical cans of two different sizes. One can is 6 inches in height and 3 inches in diameter. The second cylindrical can is 7 inches in height and 6 inches in diameter. Which size is the more economical to purchase if 4 of the smaller cans sell for the same price as one of the larger cans?

Chapter 9

GEOMETRIC RELATIONSHIPS

A Right Circular Cone

REMEMBER

Properties of a right circular cone:

Volume of a right circular cone equals one-third of the product of the area of its base and its altitude.

$$\text{VOLUME} = \frac{1}{3}(\text{BASE AREA} \times \text{ALTITUDE})$$

$$V = \frac{1}{3}Bh$$

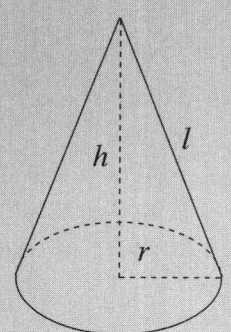

Lateral area of a right circular cone equals one-half the product of the circumference of its base and the slant height.

$$\text{LATERAL AREA} = \frac{1}{2}(\text{CIRCUMFERENCE of BASE} \times \text{THE SLANT HEIGHT})$$

$$S = \frac{1}{2}(2\pi r l) \qquad \text{[where r is the radius and } l \text{ is the slant height]}$$

Example 1: A right circular cone has an altitude of 3 ft and a base with a radius of 4 ft. Find the volume, V, to the nearest cu ft and the lateral area, S, of the cone to the nearest sq ft.

Solution 1: $V = \frac{1}{3}Bh = \frac{1}{3}\pi r^2 h = \frac{1}{3}\pi(4)(4)(3) = 50$ cu ft   Answer

$S = \frac{1}{2}(2\pi r l)$

The slant height, l, is the hypotenuse of a right triangle with legs of length r and length h.
$l^2 = r^2 + h^2$
$l^2 = 4^2 + 3^2 \quad \therefore \quad l = 5$ ft

$l = 5$ ft. $S = \frac{1}{2}(2\pi)(4)(5) = 20\pi = 63$ sq ft   Answer

Example 2: Find the volume of a right circular cone when the radius of the base and the altitude of the cone are each 12 cm in length. Round to the nearest cu cm

Solution 2: $V = \frac{1}{3}Bh = \frac{1}{3}\pi r^2 h = \frac{1}{3}\pi(12)^2(12) = 1810$ cu cm   Answer

Example 3: How many ice cream cones 2 inches high and 1.5 inch in diameter can be completely filled from a 1 gallon container of ice cream?
[1 gallon = 231 in³]

Solution 3: $V = \frac{1}{3}Bh = \frac{1}{3}\pi r^2 h = \frac{1}{3}\pi(0.75)^2(2) = 0.375\pi$ cu in

$\frac{231}{0.375\pi} = 196$ cones per gallon   Answer

Chapter 9

GEOMETRIC RELATIONSHIPS

A Right Circular Cone (continued)

1. Find the volume of a right circular cone whose radius is 15 units and height is 32 units.

2. Find the altitude of a right circular cone whose radius is 21 cm and whose volume equals the volume of a right circular cone with a radius of 17 cm and altitude of 40 cm.

3. A right circular cone and a cylinder have the same base area and the same altitude. Find the ratio of the volume of the cylinder to the volume of the cone.

4. Find the lateral area, S, of a right circular cone whose radius is 4 inches and whose slant height is 5 inches. Round to the nearest tenth of a square inch.

5. If the total area, T, is the sum of the lateral area, S, and the base area, B, find the total area of a right circular cone whose radius is 25 units and whose slant height is 40 units.

    Answer may be left in terms of π.

    [Hint: T = S + B]

Chapter 9
GEOMETRIC RELATIONSHIPS

A Right Circular Cone (continued)

6. The volume of a right circular cone is 80 π and the radius is 4. Find the altitude.

7. The volume of a right circular cone is 48 π and its altitude is 9. Find the radius of the base.

8. Find the volume in terms of π of a right circular cone whose radius is 3 and whose altitude is 9.

9. If the section of a cone made by a plane parallel to the base is a circle, then the cone is a circular cone.

    (1) SOMETIMES   (2) ALWAYS   (3) NEVER

Chapter 9

GEOMETRIC RELATIONSHIPS

A Right Circular Cone (continued)

REMEMBER

Example:
Show the lateral area, S, of a right circular cone in terms of its altitude, h, and the radius, r, if its base can be represented as: $(\pi r)(h^2 + r^2)^{\frac{1}{2}}$

Solution:
$S = \frac{1}{2}(2\pi rl)$ [where r is the radius and l is the slant height]

The slant height, l, is the hypotenuse of a right triangle with legs h and r.
Therefore, $l = (h^2 + r^2)^{\frac{1}{2}}$

Since $S = \frac{1}{2}(2\pi rl)$

$S = \frac{1}{2}(2\pi r)(h^2 + r^2)^{\frac{1}{2}} = (\pi r)(h^2 + r^2)^{\frac{1}{2}}$ Answer

---

10. Find the lateral area of a right circular cone if the radius of its base is 4 feet and its slant height is 7 feet. Round the answer to the nearest sq ft.

---

11. A plane parallel to the base of a right circular cone cuts the altitude in a point 4 inches from the vertex. If the altitude is 10 inches and the area of the section formed is 20 square inches, find the area of the base.

    Similar right circular cones: $\dfrac{B_1}{B_2} = \dfrac{(h_1)^2}{(h_2)^2}$ [Note: B is base area, h is altitude]

---

12. Find the volume, V, of a frustum formed when a plane is past through a right circular cone and the plane is parallel to the base of the cone. The frustum has two bases. The radii of the two bases of the frustum are $r_1 = 14$ ft and $r_2 = 7$ ft. The altitude of the frustum, h, is 6 feet. Round to nearest integer.

    Use the formula: $V = \frac{1}{3}\pi h(r_1^2 + r_2^2 + r_1 r_2)$

Chapter 9

GEOMETRIC RELATIONSHIPS

A Right Circular Cone (continued)

REMEMBER

Example:
A solid has the form of a right circular cone whose radius is 3 feet and whose altitude is 4 feet. When the solid, with vertex down, is placed in water, it sinks to a depth of $2\frac{2}{3}$ feet. What is the volume of that part of the solid which is under water?

Solution:
The volume of the right circular cone: $V = \frac{1}{3}Bh = \frac{1}{3}\pi r^2 h$

$V = \frac{1}{3}\pi (3)^2(4) = 12\pi$ cu ft

Similar right circular cones: $\dfrac{V_1}{V_2} = \dfrac{(h_1)^3}{(h_2)^3}$ $\qquad \dfrac{V_1}{12\pi} = \dfrac{(2\frac{2}{3})^3}{(4)^3}$

$V_1 = 3.55\pi$ cubic feet or 11.2 cubic feet to the nearest tenth   Answer

13. Find the volume of a right circular cone if the measure of the radius is 4 and its height is 9. Answer may be left in terms of π.

14. Find the volume of a right circular cone whose height is 30 and whose base has a radius of 7. Answer may be left in terms of π.

15. The altitude of a right circular cone has a measure, h, and the radius of its base has a measure, r. If r = h, express, in terms of r, the volume of the cone.

Chapter 9

GEOMETRIC RELATIONSHIPS

A SPHERE

REMEMBER

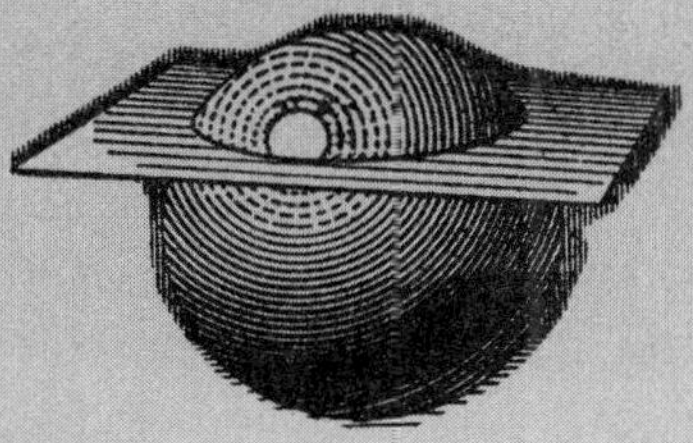

Properties of a sphere:

The intersection of a plane and a sphere is a circle.

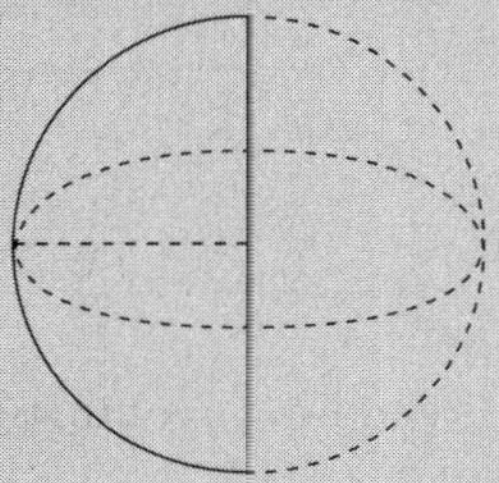

A great circle is the largest circle that can be drawn on a sphere.
A great circle of a sphere is a circle whose plane passes through the center of the sphere.

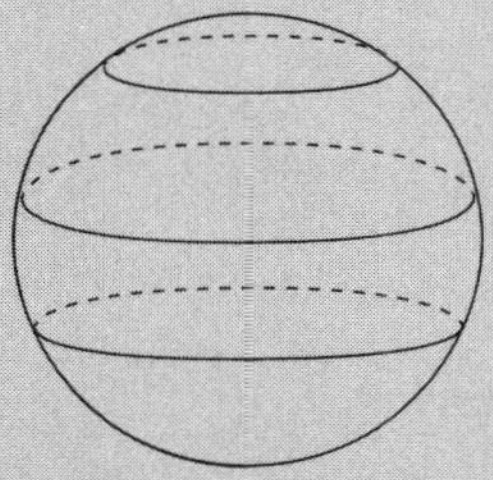

Two planes equidistant from the center of the sphere and intersecting the sphere do so in congruent circles.

Volume of a sphere equals $\dfrac{4\pi r^3}{3}$

Surface area of a sphere equals $4\pi r^2$

Example 1: If the area of a great circle of a sphere is 20, find the area of the sphere.

Solution 1: The radius of a great circle is equal to the radius of the sphere.
$A = \pi r^2$ where A is the area of the great circle
$20 = \pi r^2$

$r = \sqrt{\dfrac{20}{\pi}}$

Surface area of a sphere equals $4\pi r^2$

$S = 4\pi r^2 = 4\pi(\dfrac{20}{\pi}) = 80$ Answer

Example 2: Find the volume of the sphere in example 1 above to the nearest integer.

Solution 2: Volume of a sphere equals $\dfrac{4\pi r^3}{3} = \dfrac{4\pi(\dfrac{20}{\pi})^{\frac{3}{2}}}{3} = 67$ Answer

201

Chapter 9

GEOMETRIC RELATIONSHIPS

A SPHERE (continued)

1.	A sphere has a radius of measure 10. Find the area of a section of this sphere made by a plane at a distance 6 from the center.
2.	If a sphere having a radius of measure of 17 is cut by a plane whose distance from the center of the sphere is 15, find the measure of the radius of the circle made by the intersection of the plane and the sphere. [answer to nearest integer]
3.	What is the volume, nearest integer, of a sphere having a diameter of measure 6?
4.	Calculate the surface area of a sphere whose radius has a measure of 12. [answer to nearest integer]
5.	The radii of two concentric spheres are 15 and 17. If a plane tangent to one sphere intersects the other, find the radius of the circle of intersection.

REMEMBER

Example:
What is the measure of the radius of a sphere if the surface area of the sphere is equal to its volume?

Solution:
$S = 4\pi r^2$; $V = \dfrac{4\pi r^3}{3}$; $4\pi r^2 = \dfrac{4\pi r^3}{3}$; $3 = r$; $r = 3$ Answer

6.	The length of the radius of a sphere is 17. Find the length of the radius of the intersection of the sphere and a plane which is 8 units from the center of sphere.
7.	A circular right cylinder and a sphere have radii of equal measure. If their volumes are also equal, find the ratio h:r where h is the height of the cylinder and r is the length of the radius.

202

Chapter 9
GEOMETRIC RELATIONSHIPS

A SPHERE (continued)

8.	A plane intersects a sphere of radius 10. The intersection of the plane and the sphere has radius 6. Find the distance of the plane from the center of the sphere.
9.	Find the radius of a sphere whose surface area is 100π.
10.	A sphere of radius 1 has its volume equal to the volume of a right circular cylinder whose altitude is 3. Find the length of the radius of the cylinder.
11.	Find the area of a circle, in terms of π, formed by passing a plane 5 inches from the center of a sphere whose radius is 13 inches.
12.	Express in terms of π the volume of a sphere whose radius is 6.
13.	The volume of a sphere is $\dfrac{500\pi}{3}$. Find the radius of the sphere.
14.	All great circles of the same sphere bisect each other. *True or False?*

Chapter 9

GEOMETRIC RELATIONSHIPS

A SPHERE (continued)

REMEMBER

Example:
A solid spherical ball 4 inches in diameter weighs 10 pounds. Find the weight of a second solid spherical ball of the same material 8 inches in diameter.

Solution:
Volume of a sphere equals $\frac{4\pi r^3}{3}$

$$V_1 = \frac{4\pi r^3}{3} = \frac{4\pi(2)^3}{3} = \frac{32\pi}{3} \qquad V_2 = \frac{4\pi r^3}{3} = \frac{4\pi(4)^3}{3} = \frac{256\pi}{3}$$

$V_1 : V_2 = \text{weight}_1 : \text{weight}_2$

$\frac{32\pi}{3} : \frac{286\pi}{3} = 10 : \text{weight}_2 \qquad\qquad \text{weight}_2 = 80 \text{ pounds} \qquad\text{Answer}$

---

15. Find the radius of the sphere whose volume is 36π.

16. Find the volume, in terms of π, of a sphere whose radius is 3.

17. The radius of a small circle of a sphere is 3 inches and the plane of the circle is 4 inches from the center of the sphere. Find the area, in terms of π, of the sphere.

18. The area of a sphere is equal to the area of how many great circles of the same sphere?

Chapter 9

GEOMETRIC RELATIONSHIPS

A SPHERE (continued)

REMEMBER

The radius of a sphere circumscribed about a cube is $\frac{1}{2}$ the diagonal of the cube.

Example:

A sphere is circumscribed about a cube whose total area is 96. Find the volume of the sphere.

Solution:

Total area of a cube: $K = 6s^2$
$$96 = 6s^2$$
$$s = 4$$

Diagonal of the square: $s^2 + \sqrt{s^2} = $ (diagonal of the square)^2
$$s\sqrt{2} = \text{diagonal of the square}$$

Diagonal of cube: (Diagonal of the square)^2 + s^2 = (diagonal of the cube)^2

$$2s^2 + s^2 = \text{(diagonal of the cube)}^2$$

$$3s^2 = \text{(diagonal of the cube)}^2$$

$$3(4)^2 = 48 = \text{(diagonal of the cube)}^2$$

Diagonal of the cube $= \sqrt{48} = 4\sqrt{3}$

The radius of a sphere circumscribed about a cube is $\frac{1}{2}$ the diagonal of the cube.

$$r = \frac{1}{2}(4\sqrt{3}) = 2\sqrt{3}$$

$$V = \frac{4\pi r^3}{3} = \frac{4\pi(2\sqrt{3})^3}{3} = 32\pi\sqrt{3} = 174 \text{ cu units to nearest integer} \qquad \text{Answer}$$

Chapter 9

GEOMETRIC RELATIONSHIPS

A SPHERE (continued)

19.	The radius of a sphere is 13 inches. Find the area of a small circle whose plane is 5 inches from the center of the sphere. [find answer to the nearest tenth]
20.	The volume of a sphere is 122 cu in. Find, to the *nearest tenth*, the radius of the sphere.
21.	Find to the *nearest tenth*, the surface area of a sphere whose *diameter* is 6.2 inches.
22.	Find the volume of a sphere whose radius is 3. [find answer to nearest integer]
23.	The area of a great circle of a sphere is 3π. Find the surface area of the sphere to the nearest tenth.

REMEMBER

Example:

If lemon ice is bought by the gallon, find the least number of gallons that must be purchased in order to serve 350 persons if each serving is in the form of a hemisphere whose diameter is 2.1 inches.
 [1 gallon = 231 cu. in.]

Solution:

$$V = \frac{4\pi r^3}{3} = \frac{4\pi(1.05)^3}{3}$$

$$\frac{V}{2} = \frac{2\pi(1.05)^3}{3} \qquad \text{since each serving is a hemisphere}$$

$$\frac{350(V)}{2} = \frac{350(2\pi)(1.05)^3}{3} = 808.1747101... \text{ cu in} = 3.498591819...\text{gallons}$$

Need a minimum of 4 gallons Answer

Chapter 9

GEOMETRIC RELATIONSHIPS

A SPHERE (continued)

REMEMBER

Example 1:
A sphere is circumscribed about a rectangular solid. If the volume of the rectangular solid is 972 cubic inches and the edges are in the ratio 2:3:6, find the diameter of the sphere to the nearest inch. [The diameter of the sphere = the diagonal of the rectangular solid]

Solution 1: Volume of rectangular solid: $972 = 2x(3x)(6x)$; $x^3 = 27$; $x = 3$ inches
The dimensions of the rectangular solid are: 6 in., 9 in., and 18 in.
The diagonal of a face of the rectangular solid:
$\text{diag}_1^2 = 6^2 + 18^2$
The diagonal of the rectangular solid:
$\text{diag}_2^2 = \text{diag}_1^2 + 9^2 = 6^2 + 18^2 + 9^2$
$\text{diag}_2^2 = 441$
diameter of the sphere $= \sqrt{411} = 21$ inches   Answer

Example 2:
If the diameter of a sphere is 21 inches, find the volume of the sphere to the nearest cu in.

Solution 2:

radius of the sphere $= \dfrac{21}{2} = 10.5$ inches

$V = \dfrac{4\pi r^3}{3} = \dfrac{4\pi(10.5)^3}{3} = 4849$ cubic inches   Answer

Example 3:
3 inch diameter oranges sell at: 3 for $ 1.00
Find the selling price of ; 4 inch diameter oranges.
All oranges are of the same quality and are assumed to be spherical.

Solutions 3:

$V_3 = \dfrac{4\pi r^3}{3} = \dfrac{4\pi(1.5)^3}{3}$; $V_4 = \dfrac{4\pi r^3}{3} = \dfrac{4\pi(2)^3}{3}$

$V_3 = 14.13716694...$; $V_4 = 33.51032164...$

Cost each $= \$0.33\dfrac{1}{3}$; cost each $= x$

The cost of the 4 inch oranges should be $0.79 each   Answer

Chapter 9

GEOMETRIC RELATIONSHIPS

Pythagorean Theorem with diagonals of rectangular solids

REMEMBER

In a right triangle, the square of the hypotenuse is equal to the sum of the squares of its sides.

Hypotenuse^2 = (side$_1$)^2 + (side$_2$)^2 or $c^2 = a^2 + b^2$

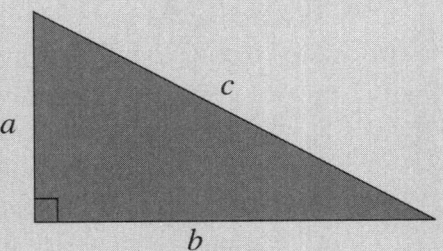

Pythagorean Triples:	3:4:5	5:12:13	8:15:17
[Multiplied by 2]	6:8:10	10:24:26	16:30:34
[Multiplied by 3]	9:12:15	15:36:39	24:45:51
[Multiplied by 10]	30:40:50	50:120:130	80:150:170

Example: If the dimensions of a rectangular solid are 4, 4 and 7 units, what is the measure of a diagonal of the solid?

Solutions: First, find the diagonal of any side: hypotenuse^2 = (side$_1$)^2 + (side$_2$)^2

(Diagonal of a side)^2 = 4^2 + 4^2

(Diagonal of a side)^2 = 32

Diagonal of a side = $\sqrt{32}$

Then, find the diagonal of the rectangular solid using the third side:

(Diagonal of the solid)^2 = (Diagonal of a side)^2 + (Diagonal of a side)^2

(Diagonal of the solid)^2 = ($\sqrt{32}$)^2 + (7)^2

(Diagonal of the solid)^2 = 32 + 49

(Diagonal of the solid)^2 = 81

Diagonal of the solid = 9 units   Answer

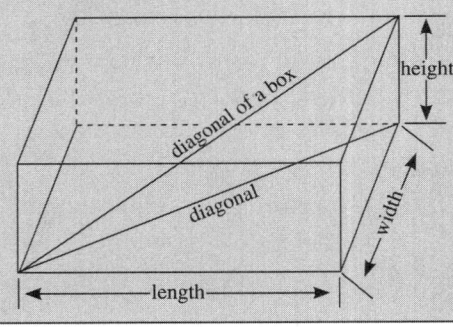

1. If the measure of the diagonal of a rectangular solid is 15 and two dimensions of the solid are 10 and 10, find the measure of the third dimension of the solid.

2. Find the length, to the **nearest tenth** of **an inch**, of the diagonal of a cube if the measure of one of the edges is 7 inches.

Chapter 9

GEOMETRIC RELATIONSHIPS

Pythagorean Theorem with diagonals of rectangular solids (continued)

3. The dimensions of a rectangular solid are 3, 4 and 12 cm. Find the length of the diagonal of the rectangular solid.

4. A set of all segments have their endpoints as points of a rectangular solid. What is the length of the longest of these segments if the dimensions of the rectangular solid are 4 ft, 6 ft and 12 ft?

5. Find the length, to the *nearest tenth* of an inch, of a diagonal of a cube whose edge is 10 inches in length.

6. If the diagonal of a face of a cube has a length of 4 cm, find the length of the diagonal of the cube to the nearest tenth of a cm.

7. Find the edge of a cube if its diagonal is $4\sqrt{3}$ ft.

8. A diagonal of a cube is $a\sqrt{3}$. Express the volume of the cube in terms of a.

Chapter 9

GEOMETRIC RELATIONSHIPS

Pythagorean Theorem with diagonals of rectangular solids (continued)

9. Find the diagonal of a rectangular solid whose dimensions are 2 ft, 6 ft and 9 ft.

10. A diagonal of a cube is $5\sqrt{3}$ units. Find the length of an edge.

11. A cube is inscribed in a sphere whose diameter is d units. Express the edge of the cube in terms of d to the *nearest thousandth*.

12. Find the length of a diagonal of a cube whose total surface area of its six sides is 24 square inches. Round answer to *nearest hundredth of an inch*.

13. If the total area of the six sides of a cube is 150 cm², find the diagonal of the cube to the *nearest tenth of a cm*.

Chapter 9

GEOMETRIC RELATIONSHIPS

Ratios in similar right circular cylinders & similar right circular cones

REMEMBER

In similar right circular *cylinders*,
  the ratio of their lateral areas = the ratio of the *squares* of their altitudes

  the ratio of their lateral areas = the ratio of the *squares* of the radii of their bases

  the ratio of their volumes = the ratio of the *cubes* of their altitudes

  the ratio of their volumes = the ratio of the *cubes* of the radii of their bases

In similar right circular *cones*,
  the ratio of their lateral areas = the ratio of the *squares* of their slant heights

  the ratio of their lateral areas = the ratio of the *squares* of their altitudes

  the ratio of their lateral areas = the ratio of the *squares* of the radii of their bases

  the ratio of their volumes = the ratio of the *cubes* of their altitudes

  the ratio of their volumes = the ratio of the *cubes* of the radii of their bases

Example: The ratio of the lateral areas of two similar right circular cylinders is 4:9. What is the ratio of the volume of the smaller cylinder to that of the larger?

Solution: Since ratio of areas is 4:9, ratio of altitudes is 2:3. Therefore ratio of volumes is 8:27   Answer

---

1. The ratio of the volume of two similar right circular cones is 27:64. Find the ratio of the lateral areas of the smaller cone to the larger cone.

---

2. What is the ratio between the areas of two similar right circular cylinders if the length of one radius is twice the length of the other?

---

3. The ratio of the radii of two similar right circular cones is 2:5. What is the ratio of the volume of the larger cone to the volume of the smaller cone?

Chapter 9

GEOMETRIC RELATIONSHIPS

Ratios in similar right circular cylinders & similar right circular cones

4. Two tanks in the form of right circular cylinders have their corresponding dimensions in the ratio 3:4. If 9 gallons of paint were needed to paint the smaller tank, how many gallons would be needed to paint the larger tank?

5. Two plastic funnels in the form of right circular cones have their corresponding dimensions in the ratio 2:5. If 1500 ml of material are needed to manufacture the larger funnel, how many ml of material would be required to manufacture the smaller funnel?

CHAPTER 10

USING A GRAPHING CALCULATOR

Solving Equations Using SOLVER

> **REMEMBER**
>
> To solve an equation that has been set equal to zero, use the SOLVER feature of your calculator.
>
> **Example:** Solve for the positive root of $x^2 - 5x - 6 = 0$.
>
> **Answer:** (1) Press the MATH button and then scroll down to #0: Solver and press ENTER.
> (2) Type in the equation $0 = x^2 - 5x - 6$.
> (3) Press ALPHA and the SOLVE to obtain the answer $x = 6$.
> Always check your answer to see if it is correct.
>
> Screen shots of how the calculator screen will appear are shown below.
>
>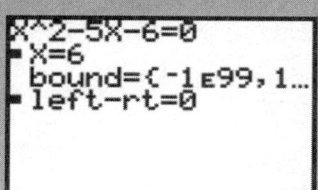

1. In the diagram below, $\triangle ABC$ is a right triangle with right angle B. $\overline{BD}$ is the altitude drawn from $\angle ABC$ to $\overline{AC}$. Determine the value of $\overline{BD}$ to the nearest hundredth.

   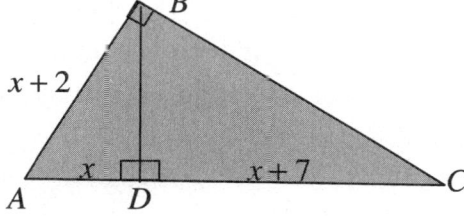

2. A sphere and a right circular cylinder have the same radius and the same volume. If the height of the cylinder is 5 inches, which of the following is the radius of the sphere?

   (1) π inches (3) 3 inches
   (2) 3.25 inches (4) 3.75 inches

3. The accompanying diagram represents a target comprised of three concentric circles. The radius of the inner circle is x. Determine a value for x such that the area of the shaded region is 52.

   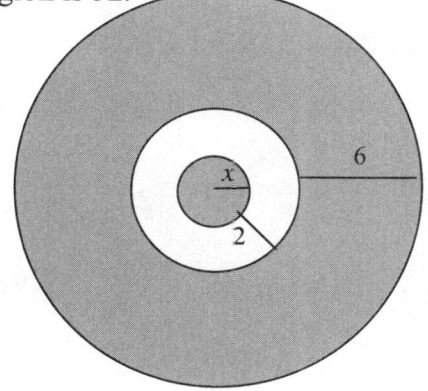

4. For which value of the radius of a sphere will the numerical value of the volume equal the numerical value of the surface area?

   (1) 1 (3) 3
   (2) 2 (4) 4

CHAPTER 10

USING A GRAPHING CALCULATOR

Solving Equations by Determining the Zeros of a Function

REMEMBER

To solve an equation by determining the zeros of a function, type the function into Y_1 of your calculator and, using the CALC menu, determine the zeros.

Example: Solve $3x^2 - 5x - 3 = 2x^2 - 3x$ for the positive value of x.

Answer:
(1) Rewrite the equation so the equation is set equal to zero. $3x^2 - 5x - 3 - (2x^2 - 3x) = 0$
(2) Type the following into Y_1 of your calculator: $Y_1 = 3x \wedge 2 - 5x - 3 - (2x \wedge 2 - 3x)$
(3) Graph Y_1. Press 2nd CALC, scroll to #2:ZERO and press ENTER.
(4) Trace to a left-bound point, press ENTER, trace to a right-bound point, press ENTER and then trace to a point close to the zero (or x-intercept) and press ENTER. The solution to the equation will appear in the lower left corner of the calculator screen. ($x = 3$)

Some screen shots of the above solution are below.

1. Use the information given in the picture below to determine the measure of $\angle DBC$.

   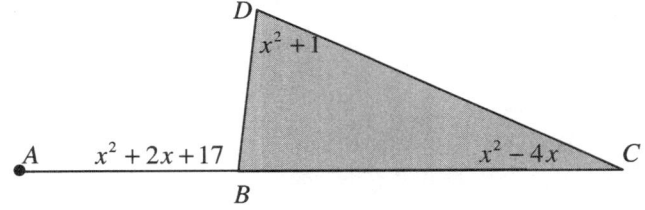

2. In right triangle CTH, $\overline{AY} \parallel \overline{TH}$. Use the information given in the picture to determine the length of $\overline{AY}$ to the nearest hundredth.

   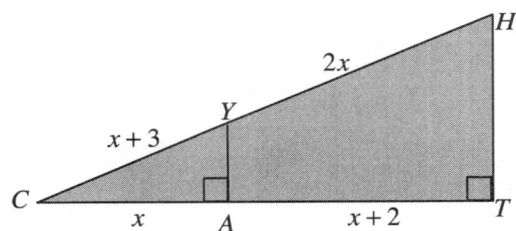

3. The Great Pyramid of Giza is a right pyramid with a square base. The measurements include a base, b, equal to 755 feet and a slant height, s, equal to 611 feet. If the ancient Egyptians had constructed a cylinder with the same volume and height as the Great Pyramid of Giza, what would be the value of the radius?

   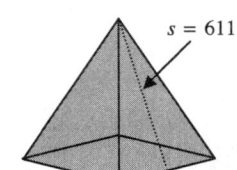

CHAPTER 10

USING A GRAPHING CALCULATOR

Graphing Polygons Using Scatterplots

> **REMEMBER**
>
> To graph a polygon, put the x-coordinates in List1 and the y-coordinates in List2. Be sure to put the first point of the polygon as the last point of the polygon.
>
> **Example:** Graph the quadrilateral formed by the points $A(3, 1)$, $B(0, 4)$, $C(-3, 2)$ and $D(-3, -3)$.
>
> **Answer:**
> (1) Turn off or delete any equations typed in the y= menu.
> (2) Press the STAT button and then choose #1:Edit.
> (3) Type the points 3, 0, −3, −3 and 3 into L1 and the points 1, 4, 2, −3 and 1 into L2.
> (4) Press 2nd STAT PLOT and choose #1: Plot 1.
> (5) Turn the plot ON and choose the second type of graph. Be sure the XList is set to L1 and the YList is set to L2.
> (6) Press ZOOM and the choose #9:ZoomStat.
> (7) Press ZOOM and the choose #5:ZSquare to get a true representation of the polygon.
>
> Screen shots of the steps above are shown below:
>
>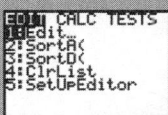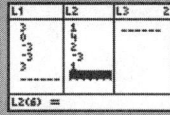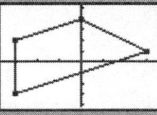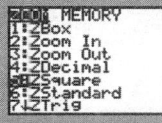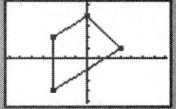
>
> Step (2)   Step (3)   Step (4)   Step (5)   Step (6)   Step (7)   Step (8)

1. Graph the quadrilateral formed by the points $A(5, 2)$, $B(1, 5)$, $C(-3, 5)$ and $D(5, -1)$. If sides $\overline{BC}$ and $\overline{DA}$ are extended, what type of angle will be formed by their intersection?

2. Graph the triangle formed by the points $A(-7, 4)$, $B(-2, -5)$ and $C(6, -7)$. What type of triangle is $\triangle ABC$?

3. A pyramid with a rectangular base is to be constructed so that the vertices of the base, when graphed on a coordinate axis are at the points $A(4, 1)$, $B(1, 4)$, $C(-8, -5)$ and $D(-5, -8)$. At what point on the coordinate axis should the pyramid's height begin so that the distance from the vertex of the pyramid to each point on the rectangular base is the same?

4. A pyramid with a triangular base is to be constructed so that the vertices of the base, when graphed on a coordinate axis are at the points $A(-7, 8)$, $B(-1, -4)$ and $C(3, 8)$. At what point on the coordinate axis should the pyramid's height begin so that the distance from the vertex of the pyramid to each point on the triangular base is the same?

215

CHAPTER 10

USING A GRAPHING CALCULATOR

Using Scatterplots and Lists to Transform Polygons

> **REMEMBER**
>
> To determine the coordinate points and graph of a polygon after a transformation use List1, List2, List3 and List4. The polygon's original points are put in List1 and List2. By typing the transformation for each variable and storing the result in List3 and List4, polygon and its' transformation can be graphed.
>
> **Example**: The coordinates of $\triangle DEF$ are $D(-3, 1)$, $E(-2, 6)$ and $F(1, 4)$.
>
>   (a)   Graph $\triangle D'E'F'$ the image of $\triangle DEF$ after the translation $(x, y) \rightarrow (x+10, y-7)$.
>
>   (b)   Graph $\triangle D''E''F''$ the image of $\triangle DEF$ after the dilation D_3.
>
> **Answer**:   (1)   Type the x-coordinates of $\triangle DEF$ into L1 and the y-coordinates of $\triangle DEF$ into L2.
>
>   (2)   On the HOME SCREEN, type in the rule for the x-coordinate translation and store the answer in L3. Type in the rule for the y-coordinate translation and store the answer in L4.
>
>   (3)   On the HOME SCREEN, type in the rule for the x-coordinate dilation and store the answer in L5. Type in the rule for the y-coordinate dilation and store the answer in L6.
>
>   (4)   Create and turn on three scatterplots, one for L1, L2, one for L3, L4 and one for L5, L6. (Remember to first do a ZOOM STAT and then a ZOOM SQUARE.)
>
> Screen shots of the above rules are shown below:
>
>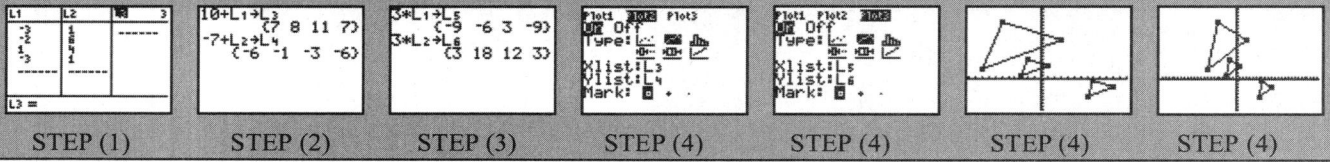
>
> STEP (1)   STEP (2)   STEP (3)   STEP (4)   STEP (4)   STEP (4)   STEP (4)

1. The coordinates of $\triangle ABC$ are $A(-8, 13)$, $B(-3, -9)$ and $C(8, 3)$.

    Graph and state the coordinates of $\triangle A'B'C'$, the image of $\triangle ABC$ after the transformation $T_{-3, 5}$.

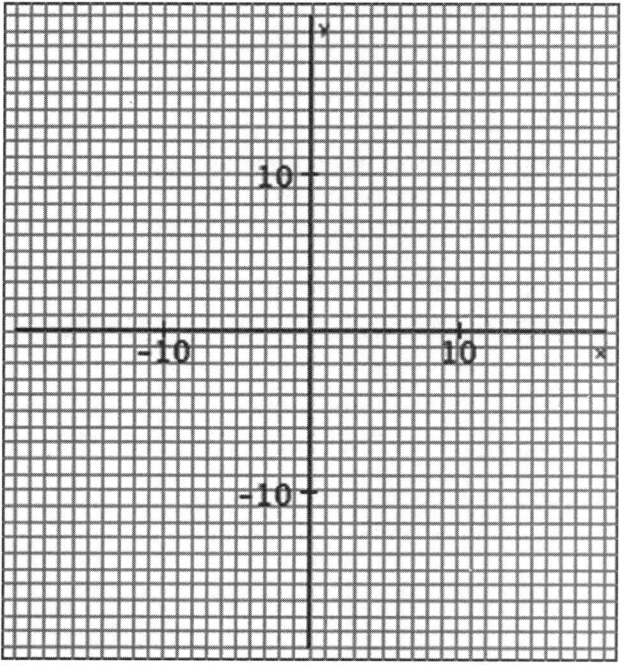

216

CHAPTER 10

USING A GRAPHING CALCULATOR

Using Scatterplots and Lists to Transform Polygons, Continued

2. The coordinates of $\triangle ABC$ are $A(-2, 1)$, $B(0,-3)$ and $C(3,4)$.

   Graph and state the coordinates of $\triangle A'B'C'$, the image of $\triangle ABC$ after the dilation D_2.

   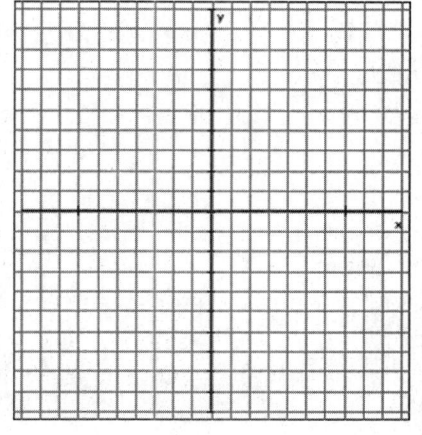

3. The coordinates of quadrilateral $JENA$ are $J(-6, 2)$, $E(-10, -6)$, $N(-6, -8)$ and $A(0, -4)$.

   Graph and state the coordinates of quadrilateral $J'E'N'A'$, the image of $JENA$ after a translation that maps $(x, y) \to (x+8, y+6)$.

   Graph and state the coordinates of quadrilateral $J''E''N''A''$, the image of $JENA$ after the dilation $D_{\frac{1}{2}}$.

   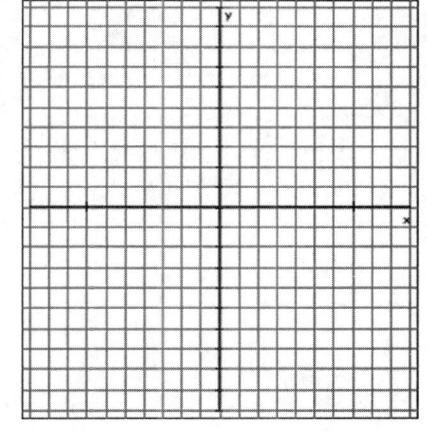

4. The coordinates of pentagon $GRACE$ are $G(-7, 3)$, $R(-9, -3)$, $A(-7, -7)$, $C(-3, -5)$ and $E(-5, -5)$.

   Graph and state the coordinates of pentagon $G'R'A'C'E'$, the image of pentagon $GRACE$, after the translation $T_{7, 1}$.

   Graph and state the coordinates of pentagon $G''R''A''C''E''$, the image of pentagon $G'R'A'C'E'$ after the dilation $D_{\frac{1}{2}}$.

   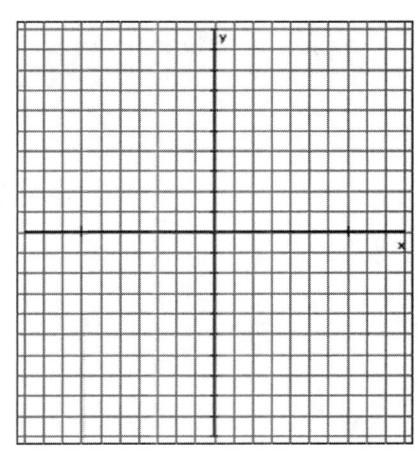

217

CHAPTER 10

USING A GRAPHING CALCULATOR

Using Scatterplots and Lists to Rotate Polygons About the Origin

REMEMBER

To the right are the standard formulas for rotating a point about the origin:

Notation	Formula	Example	Answer
$R_{(O, 90°)}$	$(x,y) \to (-y, x)$	$(3, 5)$	$(-5, 3)$
$R_{(O, 180°)}$	$(x,y) \to (-x, -y)$	$(2, 1)$	$(-2, -1)$
$R_{(O, -90°)}$	$(x,y) \to (y, -x)$	$(-3, 4)$	$(4, 3)$

To determine the coordinate points and graph of a polygon after a rotation about the origin use List1, List2, List3 and List4. The polygon's original points are put in List1 and List2. By typing the transformation for each variable and storing the result in List3 and List4, if needed, the polygon and its' rotation can be graphed.

Example: The coordinates of $\triangle ABC$ are $A(2, 0)$, $B(8, 3)$ and $C(-2, 7)$. Graph and state the coordinates of $\triangle A'B'C'$ the image of $\triangle ABC$ after a 90° clockwise rotation.

Answer:
(1) Type the x-coordinates of $\triangle ABC$ into L1 and the y-coordinates of $\triangle ABC$ into L2.
(2) On the HOME SCREEN, type in the rule for the x-coordinate clockwise rotation and store the answer in L3. Type in the rule for the y-coordinate rotation and store the answer in L4.
(3) Create and turn on two scatterplots, one for L1, L2 and one for L3, L4. (Remember to first do a ZOOM STAT and then a ZOOM SQUARE.)

Screen shots of the above rules are shown below:

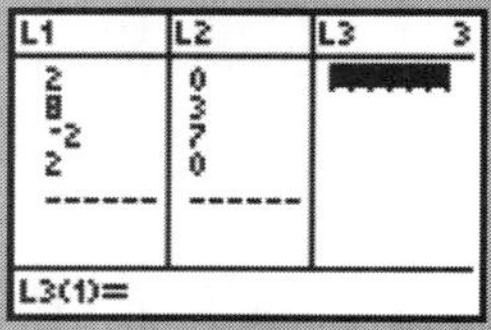

Step (1) 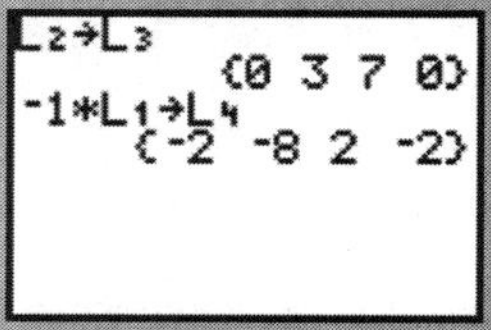   Step (2)    Step (3) 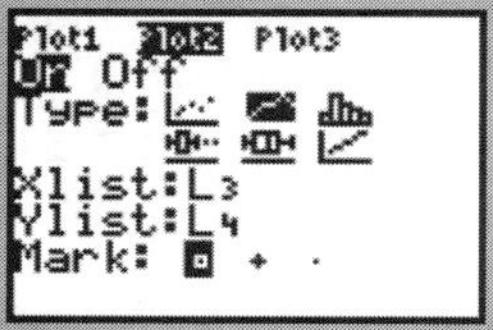   Step (3) 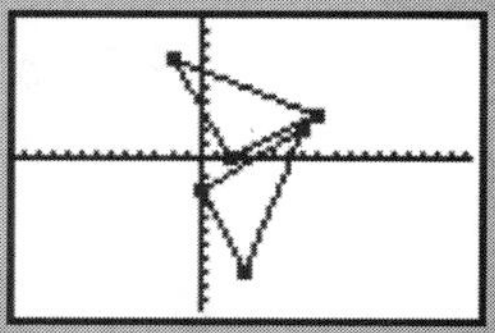   Step (3)

The coordinates of $\triangle A'B'C'$ are $A'(0, -2)$, $B'(3, -8)$ and $C(7, 2)$.

1. The coordinates of $\triangle ABC$ are $A(2, -5)$, $B(5, 3)$ and $C(-3, 5)$.

   Graph and state the coordinates of $\triangle A'B'C'$ the image of $\triangle ABC$ after a 90° counter-clockwise rotation.

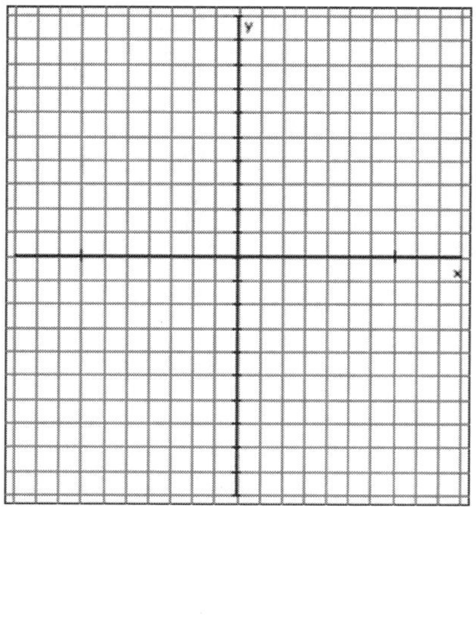

CHAPTER 10

USING A GRAPHING CALCULATOR

Using Scatterplots and Lists to Rotate Polygons About the Origin, Continued

2. The coordinates of quadrilateral $JACK$ are $J(2, 1)$, $A(6, 3)$, $C(5, 5)$ and $K(1, 7)$.

   Graph and state the coordinates of quadrilateral $J'A'C'K'$, the image of quadrilateral $JACK$, after the transformation $R_{O, 180°}$.

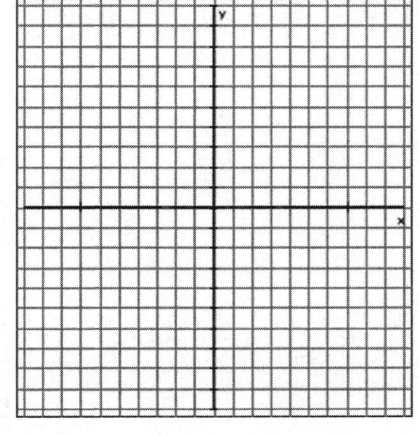

3. The coordinates of $\triangle KIM$ are $K(7, -4)$, $I(8, 7)$ and $M(3, 4)$.

   Graph and state the coordinates of $\triangle K'I'M'$, the image of $\triangle KIM$ after a 90° clockwise rotation about the origin.

   Graph and state the coordinates of $\triangle K''I''M''$, the image of $\triangle K'I'M'$ after the rotation $R_{O, 180°}$.

   Name a single transformation that maps $\triangle K''I''M''$ onto $\triangle KIM$.

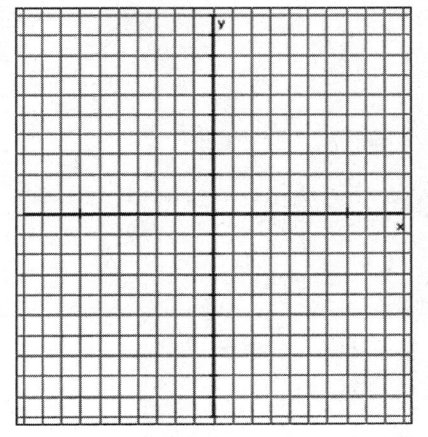

4. The coordinates of $\triangle CAS$ are $C(-1, 7)$, $A(-7, -1)$ and $S(-8, 8)$.

   Graph and state the coordinates of $\triangle C'A'S'$, the image of $\triangle CAS$ after the rotation $R_{O, -90°}$.

   Graph and state the coordinates of $\triangle C''A''S''$, the image of $\triangle CAS$ after the rotation $R_{O, 180°}$.

   Graph and state the coordinates of $\triangle C'''A'''S'''$, the image of $\triangle CAS$ after the rotation $R_{O, 90°}$.

   Name the type of quadrilateral formed by the points $CAA'A''$. Justify your answer.

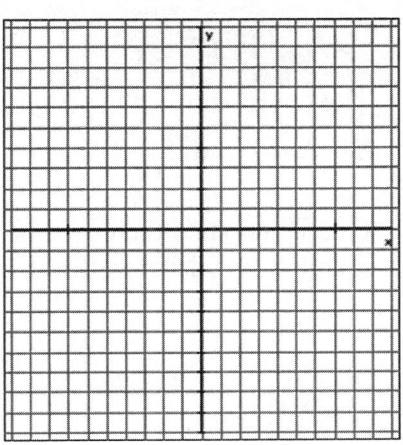

CHAPTER 10

USING A GRAPHING CALCULATOR

Using Scatterplots and Lists to Reflect Polygons Over Straight Lines

REMEMBER

To the right are the standard line reflection formulas:

Line of Reflection	Notation	Formula	Example	Answer
x-axis or $y=0$	$r_{y=0}(x,y)$	$(x,y) \rightarrow (x,-y)$	$(3,5)$	$(3,-5)$
y-axis or $x=0$	$r_{x=0}(x,y)$	$(x,y) \rightarrow (-x,y)$	$(2,1)$	$(-2,1)$
$y=x$	$r_{y=x}(x,y)$	$(x,y) \rightarrow (y,x)$	$(-3,4)$	$(4,-3)$
$y=-x$	$r_{y=-x}(x,y)$	$(x,y) \rightarrow (-y,-x)$	$(4,-2)$	$(2,-4)$

To determine the coordinate points and graph of a polygon after a line reflection use List1, List2, List3 and List4. The polygon's original points are put in List1 and List2. By typing the reflection rule for each variable and storing the result in List3 and List4, if needed, the polygon and its' reflection can be graphed.

Example: The coordinates of $\triangle ABC$ are $A(1,0)$, $B(8,3)$ and $C(3,10)$. Graph and state the coordinates of $\triangle A'B'C'$ the image of $\triangle ABC$ after a reflection in the y-axis.

Answer:
(1) Type the x-coordinates of $\triangle ABC$ into L1 and the y-coordinates of $\triangle ABC$ into L2.
(2) On the HOME SCREEN, type in the rule for the x-coordinate reflection in the y-axis and store the answer in L3. Type in the rule for the y-coordinate reflection in the y-axis and store the answer in L4.
(3) Create and turn on two scatterplots, one for L1, L2 and one for L3, L4. (Remember to first do a ZOOM STAT and then a ZOOM SQUARE.)

Screen shots of the above rules are shown below:

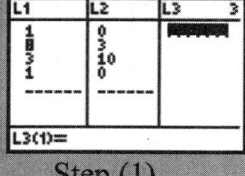 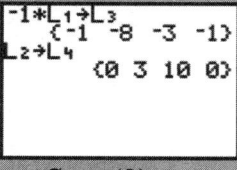   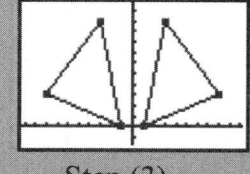

Step (1)   Step (2)   Step (3)   Step (3)   Step (3)

The coordinates of $\triangle A'B'C'$ are $A'(-1, 0)$, $B'(-8, 3)$ and $C'(-3, 10)$.

1. The coordinates of quadrilateral *MIKE* are $M(5, 7)$, $I(1, 10)$, $K(-8, 5)$ and $E(-2, 1)$.

   Graph and state the coordinates of quadrilateral $M'I'K'E'$, the image of quadrilateral *MIKE* after a reflection in the line $y = x$.

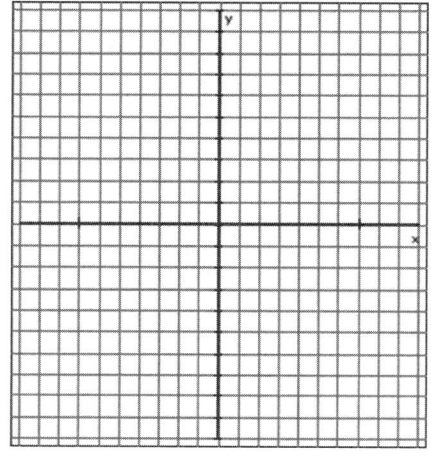

CHAPTER 10

USING A GRAPHING CALCULATOR

Using Scatterplots and Lists to Reflect Polygons Over Straight Lines, Continued.

2. The coordinates of $\triangle SAM$ are $S(-5, 6)$, $A(-8, 1)$ and $M(4, 7)$.

   Graph and state the coordinates of $\triangle S'A'M'$, the image of $\triangle SAM$, after a reflection in the line $y = 0$.

   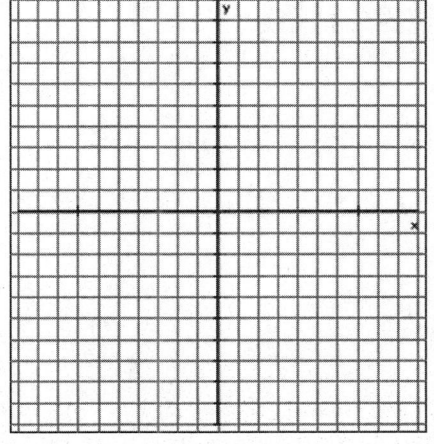

3. The coordinates of pentagon $NAGEL$ are $N(-1, -5)$, $A(2, -9)$, $G(5, -7)$, $E(5, -3)$ and $L(3, -1)$.

   Graph and state the coordinates of pentagon $N'A'G'E'L'$, the image of pentagon $NAGEL$, after a reflection in the line $y = x$.

   Graph and state the coordinates of pentagon $N''A''G''E''L''$, the image of pentagon $N'A'G'E'L'$, after a reflection in the y-axis.

   Name the single transformation that would map pentagon $NAGEL$ onto pentagon $N''A''G''E''L''$.

   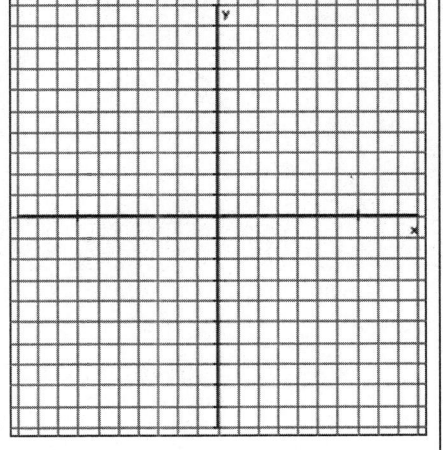

4. The coordinates of quadrilateral $UDEY$ are $U(3, 3)$, $D(7, 1)$, $E(8, 10)$ and $Y(1, 7)$.

   Graph and state the coordinates of quadrilateral $U'D'E'Y'$, the image of quadrilateral $UDEY$, after reflection in the x-axis.

   Graph and state the coordinates of quadrilateral $U''D''E''Y''$, the image of quadrilateral $U'D'E'Y'$, after a reflection in the y-axis.

   Name the single transformation that would map quadrilateral $UDEY$ onto quadrilateral $U''D''E''Y''$.

   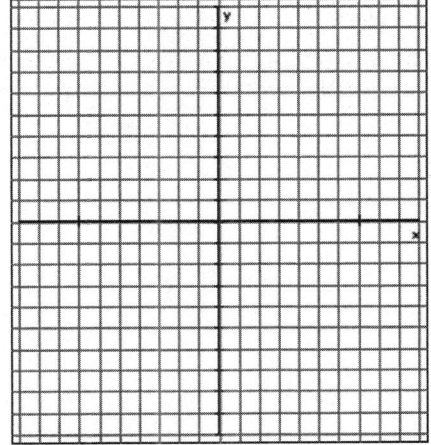

CHAPTER 10

USING A GRAPHING CALCULATOR

Graphing Circles

> **REMEMBER**
>
> Circles can be graphed from the **HOMESCREEN** of a graphing calculator. Although these circles can not be traced, having their graphs should be helpful in solving problems.
>
> **Example**: Graph the circle whose equation is $(x-3)^2 + (y+1)^2 = 25$.
>
> **Answer**:
> (1) The center of this circle is the point $(3, -1)$ and the radius is 5.
> (2) From the **HOMESCREEN** of the calculator, press **2nd DRAW** and scroll to **9: Circle(** and press **ENTER**.
> (3) The syntax for graphing a circle is Circle(Center, Radius) so type Circle(3, −1, 5) and press **ENTER**.
> (4) Adjust the graph window as necessary, but press ZOOM #5:Square to see a good representation of the circle.
>
> Screen shots of these steps are shown below:
>
>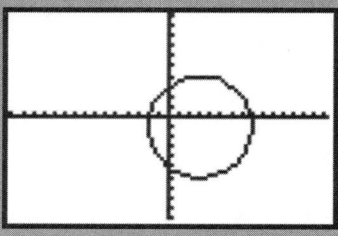
>
> Step (2)   Step (3)   Step (4)

1. Graph the circle whose center is $(2, 1)$ and whose radius is 3.

   What is the equation of this circle?

2. What are the center and radius of the circle whose equation is given by $(x+1)^2 + (y-3)^2 = 4$?

   Graph the circle on the graph paper to the right.

CHAPTER 10

USING A GRAPHING CALCULATOR

Graphing Circles, Continued

3. A circle with the equation $(x-1)^2 + (y+1)^2 = 16$ is intersected in two places by the line $y = 1$.

    Graph the circle and the line on the graph paper to the right.

    Determine the length of the chord formed by the two intersection points to the nearest hundredth. Show your work.

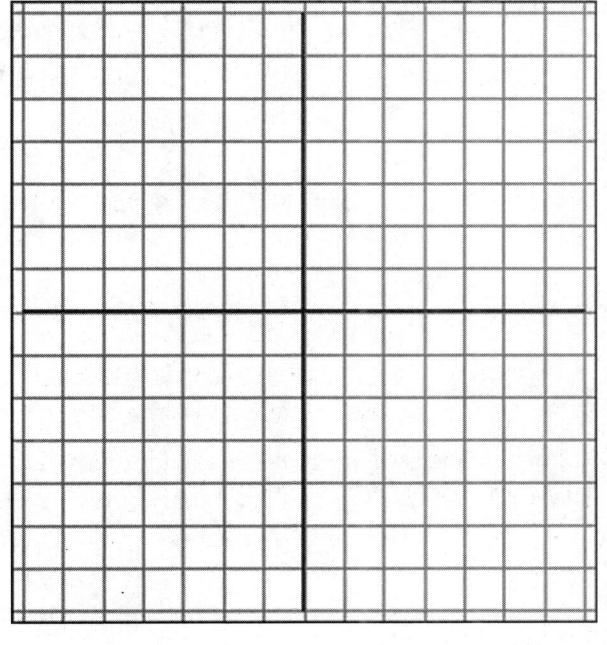

4. Graph the circle $(x+3)^2 + (y+2)^2 = 9$.

    Graph the line $2y + 3x = 4$.

    At how many points does the graph of the line intersect the graph of the circle?

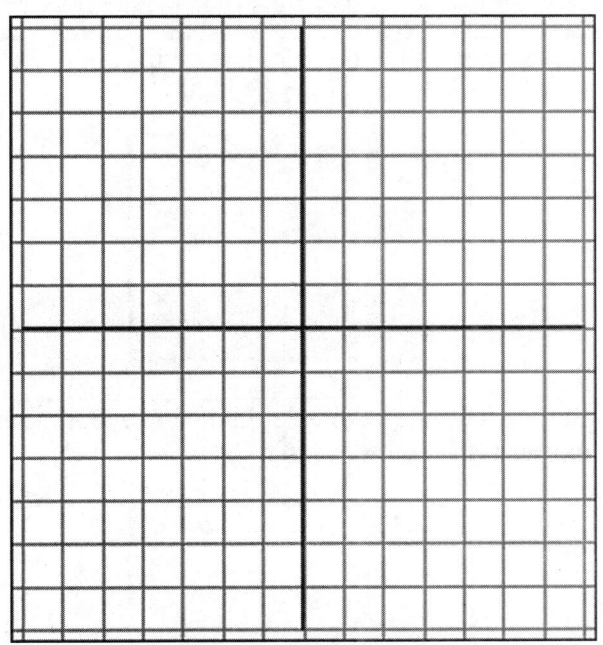

CHAPTER 10

USING A GRAPHING CALCULATOR

Know and Apply the Conditions Under Which a Compound Sentence is True

> **REMEMBER**
>
> A computer or calculator will report a value of 1 if an expression is **TRUE** and a value of 0 if an expression is **FALSE**.
>
> **Recall:** The table below summarizes the truth values for simple, compound sentences.
>
> | STATEMENTS | | NEGATIONS | | CONJUNCTION | DISJUNTION | CONDITIONAL | BICONDITIONAL |
> |---|---|---|---|---|---|---|---|
> | p | q | $\sim p$ | $\sim q$ | $p \wedge q$ | $p \vee q$ | $p \rightarrow q$ | $p \leftrightarrow q$ |
> | T | T | F | F | T | T | T | T |
> | T | F | F | T | F | T | F | F |
> | F | T | T | F | F | T | T | F |
> | F | F | T | T | F | F | T | T |
>
> **Example:** What value will a computer report for the expression shown below?
>
> `2+1=3 and 2^3=8`
>
> **Answer:** A conjunction is true when both expressions are true. Since $2 + 1$ does equal 3 and 2^3 does equal 8, this expression is true. The computer will report a value of 1.
>
> `2+1=3 and 2^3=8`
> `                1`

1. State what value the calculator will report for the expression shown in the display.

   `7-1<6 or 2*6=8`

2. State what value the calculator will report for the expression shown in the display.

   `2^2=4 and 2+2=4`

3. State what value the calculator will report for the expression shown in the display.

   `4^3=12 or 4^3=64`

4. State what value the calculator will report for the expression shown in the display.

   `2^5≤32 and 0!=0`

CHAPTER 10

USING A GRAPHING CALCULATOR

Using Geometry Software to Make Conjectures

> **REMEMBER**
> To choose a line or a point, move the cursor close to the line or point until the line or point becomes activated and some of the pixels begin to blink.
>
>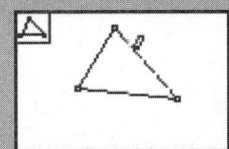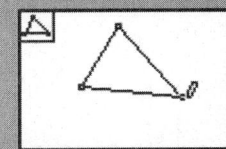
>
> **Example:** Elaina and Kyle construct $\triangle ABC$ and draw a $\overline{DE}$ connecting the midpoints of sides $\overline{AC}$ and $\overline{AB}$. They collect measurements for the lengths of $\overline{DE}$ and $\overline{BC}$ as they change the size of $\triangle ABC$ and store them in the table shown below. Write a conjecture about the relationship between the lengths of $\overline{DE}$ and $\overline{BC}$
>
>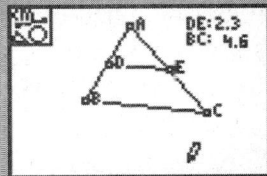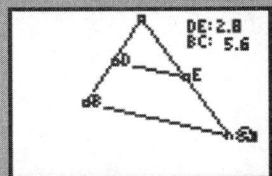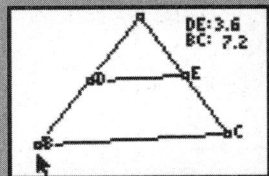
>
> | DE | BC |
> |---|---|
> | 2.3 | 4.6 |
> | 2.8 | 5.6 |
> | 3.6 | 7.2 |
>
> **Answer:** Since each length for $\overline{DE}$ is $\dfrac{1}{2}$ the length of $\overline{BC}$, the conjecture is that the length of the line segment that connects the midpoints of two sides of a triangle is equal to $\dfrac{1}{2}$ the length of the third side.

1. Using the geometry software in her calculator, Alanna drew $\triangle ABC$, overlayed a line on $\overline{BC}$ and located point D on this line as shown in the screen shot below. Alanna measured $\angle BAC$, $\angle ACB$ and $\angle ABD$ as she changed the shape of $\triangle ABC$ and recorded these measures in the table. Which of the following is a correct statement based upon her work?

   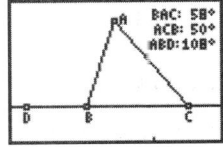 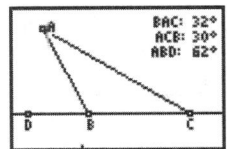 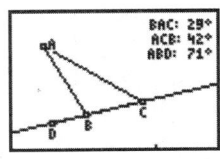

	$\triangle 1$	$\triangle 2$	$\triangle 3$
$m\angle BAC$	58°	32°	29°
$m\angle ACB$	50°	30°	42°
$m\angle ABD$	108°	62°	71°

   (1) The measure of $\angle ABC$ will always be obtuse.
   (2) The measure of an exterior angle of a triangle will equal the sum of the two nonadjacent interior angles.
   (3) The sum of the angles of a triangle drawn in a plane might not equal 180° in all cases.
   (4) The measures of the three interior angles of a triangle drawn in a plane using this software will always be less than 90°.

CHAPTER 10

USING A GRAPHING CALCULATOR

Using Geometry Software to Make Conjectures, Continued

2. Alex, using the geometry software program in his calculator, drew a pair of parallel lines $(\overleftrightarrow{EF} \parallel \overleftrightarrow{GH})$ and also drew $\overleftrightarrow{AD}$ so that it intersects $\overleftrightarrow{EF}$ at B and $\overleftrightarrow{GH}$ at C. Alex then measured $\angle GCB$ and $\angle EBC$ as he changed the location of $\overleftrightarrow{EF}$ and recorded the angle measurements in a table.

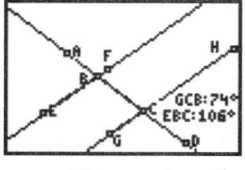
Figure 1

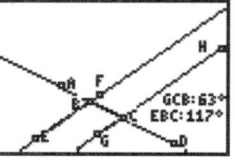
Figure 2

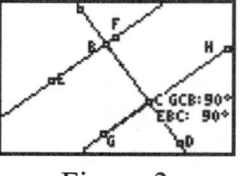
Figure 3

	$m\angle GCB$	$m\angle EBC$
Figure 1	74°	106°
Figure 2	63°	117°
Figure 3	90°	90°

Write a conjecture, based on Alex's work, about the relationship between the measures of the two interior angles on the same side of a transversal that intersects two parallel lines.

3. Kara, using the geometry software program in her calculator, drew a pair of lines $(\overleftrightarrow{AC}$ and $\overleftrightarrow{DF})$ which were both intersected by $\overleftrightarrow{GH}$ at points B and E, respectfully. Kara then measured $\angle ABE$, $\angle DEB$, $\angle CBE$ and $\angle BEF$ and recorded these angle measurements in a table.

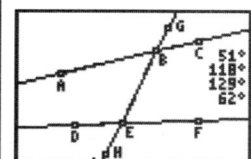
Figure 1

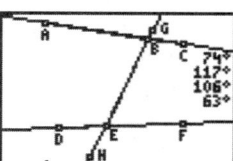
Figure 2

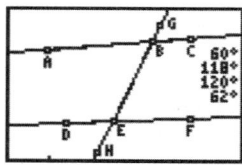
Figure 3

	Figure 1	Figure 2	Figure 3
$m\angle ABE$	51°	74°	60°
$m\angle DEB$	118°	117°	118°
$m\angle CBE$	129°	106°	120°
$m\angle BEF$	62°	63°	62°

Use Kara's work to justify the following statements:

If the alternate interior angles formed when two lines are cut by a transversal are not equal in measure then the two lines are not parallel.

If two lines are not parallel, then they will intersect on the side of a transversal in which the sum of the two interior angles is less than 180°.

CHAPTER 10

USING A GRAPHING CALCULATOR

Using Geometry Software to Make Conjectures, Continued

Use the information below to answer questions 4 - 9.

Marisa, using the geometry software program in her calculator, drew $\triangle ABC$, located the midpoint of each side of the triangle and drew line segments $\overline{AE}$, $\overline{BF}$ and $\overline{CD}$. These three lines met at a single point that she labeled P. Finally, Marisa determined the lengths of $\overline{AP}$ and $\overline{EF}$ as she changed the shape of $\triangle ABC$ and recorded these lengths in the table.

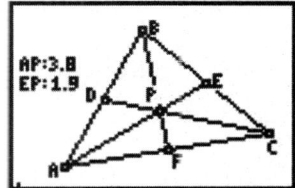
Figure 1

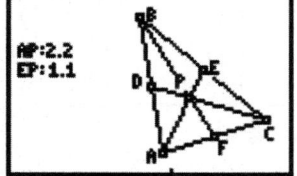
Figure 2

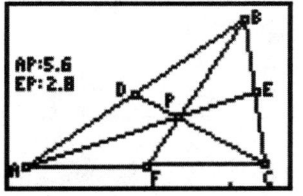
Figure 3

	AP	EF
Figure 1	3.8	1.9
Figure 2	2.2	1.1
Figure 3	5.6	2.8

4. If Marisa changed the shape of $\triangle ABC$ again, and the measure of $\overline{AP}$ was 3.4, which of the following represents the length of $\overline{EF}$?

   (1) 1.7 (3) 5.1
   (2) 3.4 (4) 6.8

5. If Marisa changed the shape of $\triangle ABC$ one more time, and the measure of $\overline{EF}$ was 2.4, which of the following represents the length of $\overline{AE}$?

   (1) 1.8 (3) 7.2
   (2) 4.8 (4) 9.6

6. Which of the following is the name for the point of intersection for the three medians of a triangle?

   (1) Incenter   (3) Circumcenter
   (2) Centroid   (4) Orthocenter

7. In Figure 3, which of the following gives the correct relationship between the area of $\triangle ABF$ and the area of $\triangle CBF$?

   (1) $A_{\triangle ABF} > A_{\triangle CBF}$ (3) $A_{\triangle ABF} = A_{\triangle CBF}$
   (2) $A_{\triangle ABF} > A_{\triangle CBF}$ (4) Can not be determined

8. Which of the following represents the ratio of $EP : EA$?

   (1) 1:2 (3) 1:3
   (2) 2:1 (4) 2:3

9. Which of the following dilations centered at point A maps point P onto point E?

   (1) $D_{\frac{1}{2}}$ (3) $D_{\frac{2}{3}}$
   (2) D_{2} (4) $D_{\frac{3}{2}}$

227

CHAPTER 10

USING A GRAPHING CALCULATOR

Using Geometry Software to Make Conjectures, Continued

10. John, using the geometry software program in his calculator, drew circle O with diameter $\overline{CD}$ perpendicular to chord $\overline{AB}$ at point E. John then measured segments $\overline{CE}$, $\overline{DE}$, $\overline{AE}$ and $\overline{BE}$ for lengths of chord $\overline{AB}$ and recorded the results in a table.

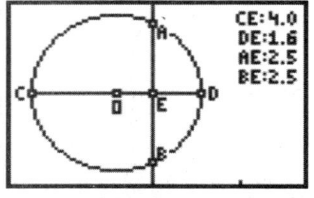
Figure 1

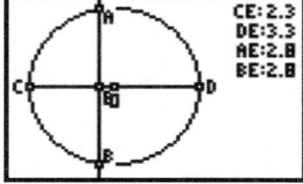
Figure 2

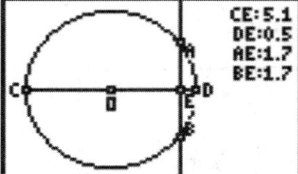
Figure 3

	Fig. 1	Fig. 2	Fig. 3
CE	4.0	2.3	5.1
DE	1.6	3.3	0.5
AE	2.5	2.8	1.7
BE	2.5	2.8	1.7

Write a conjecture based upon John's work about the relationship between the diameter of a circle and a chord that is perpendicular to the diameter.

11. Kerrin, using the geometry software program in her calculator, drew circle O with chord $\overline{CD}$ and chord $\overline{AB}$ intersecting at point E. Kerrin then measured segments $\overline{CE}$, $\overline{DE}$, $\overline{AE}$ and $\overline{BE}$ for different sized chords and recorded the results in a table.

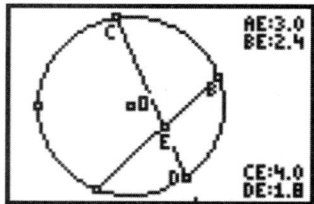
Figure 1

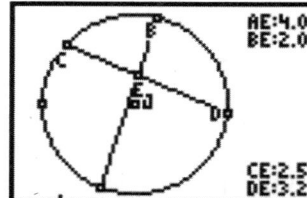
Figure 2

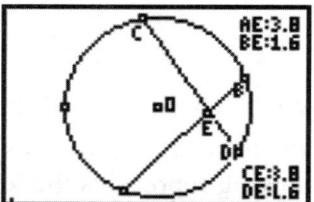
Figure 3

	Figure 1	Figure 2	Figure 3
AE	3.0	4.0	3.8
BE	2.4	2.0	1.6
CE	4.0	2.5	3.8
DE	1.8	3.2	1.6
AE × BE	7.2	8.0	6.08
CE × DE	7.2	8.0	6.08

Write a conjecture based upon Kerrin's work about the relationship between the lengths of two intersecting chords of a circle.

CHAPTER 10

USING A GRAPHING CALCULATOR

Using Geometry Software to Make Conjectures, Continued

12. Genni, using the geometry software program in his calculator, drew circle O with chord $\overline{AB}$. After locating points C, D and E on circle O, she measured $\angle ACB$, $\angle ADB$ and $\angle AEB$, as shown in the screen shots below.

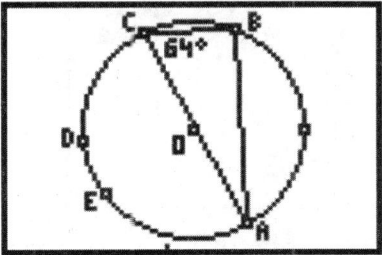 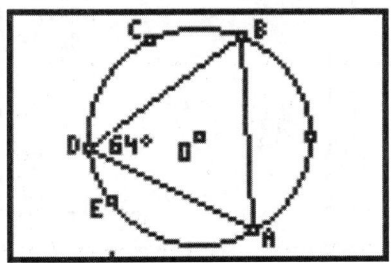 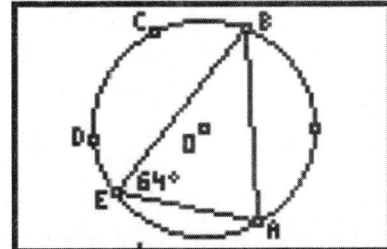

Using Genni's work, write a conjecture about the measure of different inscribed angles whose vertices are on the same side of the circle as the chord they intersect.

13. Genni then located point F on the opposite side of chord $\overline{AB}$. At first Genni thought that $\angle AFB$, shown below, would also measure 64°, until she realized that $\angle AFB$ was obtuse.

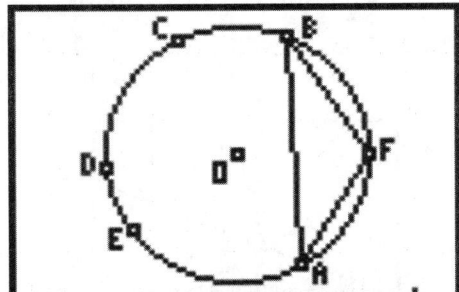

Genni knew that the measure of $\angle AFB$ would have to be related to the measures of the other inscribed angles that met chord $\overline{AB}$ from the other side of circle O.

Write a conjecture about the measure of $\angle AFB$.

Write a conjecture about the relationship between the measures of inscribed angles to the same chord whose vertices are on opposite sides of the chord.

CHAPTER 10

USING A GRAPHING CALCULATOR

Using Geometry Software to Make Conjectures, Continued

14. Jesse, using the geometry software program in his calculator, drew circle O and chord $\overline{AB}$. After locating point C on circle O, Jesse measured both $\angle AOB$ and $\angle ACB$. He did this for three different chord $\overline{AB}$s and recorded his data in a table.

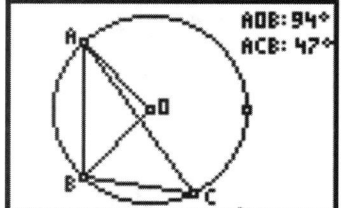 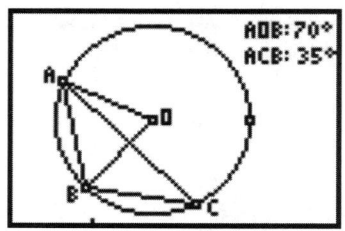 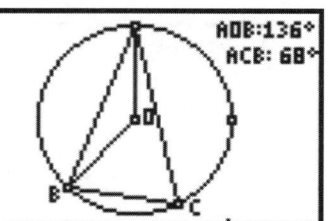

Figure 1   Figure 2   Figure 3

	Figure 1	Figure 2	Figure 3
$m\angle AOB$	94°	70°	136°
$m\angle ACB$	47°	35°	68°

Write a conjecture based on Jesse's work about the relationship between the measure of a central angle and the measure of an inscribed angle of a circle that intersect the same chord.

15. Collin, using the geometry software program in his calculator, drew $\triangle ABC$. The three altitudes, $\overline{AD}$, $\overline{BE}$ and $\overline{CF}$ meet at the orthocenter, labeled D. As Collin changed the shape of $\triangle ABC$, he noticed that the point D moved from inside the triangle to outside the triangle.

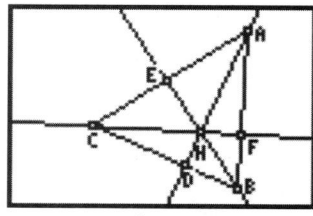 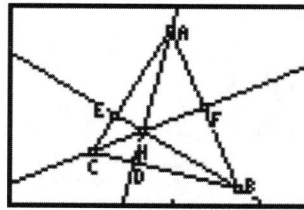 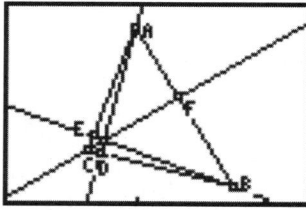

Figure 1   Figure 2   Figure 3

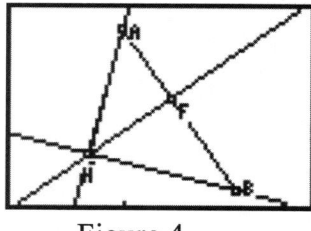 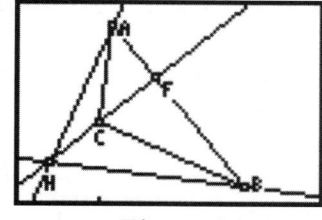

Figure 4   Figure 5

Write a conjecture, based on Collin's work, that describes the location of the orthocenter in acute, obtuse and right triangles.

Reference Sheet

The Regents Examination in Geometry will include a reference sheet containing the formulas specified below.

Volume	Cylinder	$V = Bh$ where B is the area of the base
	Pyramid	$V = \dfrac{1}{3}Bh$ where B is the area of the base
	Right Circular Cone	$V = \dfrac{1}{3}Bh$ where B is the area of the base
	Sphere	$V = \dfrac{4}{3}\pi r^3$

Lateral Area (L)	Right Circular Cylinder	$L = 2\pi rh$
	Right Circular Cone	$L = \pi rl$ where l is the slant height

Surface Area	Sphere	$SA = 4\pi r^2$

INDEX

AAA ≅ AAA and AA ≅ AA [similar triangles] 30, 31, 138, 146
AAS ≅ AAS 16
Alternate interior angles 43-45
"AND" Conjunction 5-6
Angle measurement, invariant 127, 142, 146
Angles in circles 64-65
Angles of a polygon, sum of interior and exterior 46
Angles of a regular polygon, each interior and exterior 47
Angles of a triangle, sum of interior and exterior 36
Arcs of a circle 58-71
    Parallel lines intersecting a circle 58-59
    Tangent and a chord 61, 64, 65
    Two chords intersecting in a circle 58-59
Area, invariant 127, 142, 146
    Two tangents, two secants, tangent and secant 63-66
ASA ≅ ASA 16
Biconditional 10
Bisector of vertex angle of isosceles triangle 39
Center of dilation 112
Central angles 60
Centroid of a triangle, finding coordinates 96
Centroid of a triangle, median segments 2:1 ratio 50
Circle 87-93, 141, 155-157
Chords of a circle, perpendicular bisector, distance from center 58-60
Colinearity invariant 107, 127, 142, 146
Common external and internal tangent lines to a circle 70-71
Compositions of isometries and dilations 116-122
Compound statements 14, 224
Conditional statements 9
Congruence of triangles, SSS, SAS, ASA, AAS, HYP-LEG 16-18
Conjunctions 5-6
Constructions 158-163
Constructions, centroid of a triangle 162
Constructions, circumcenter of a triangle 162
Constructions, incenter of a triangle 162
Constructions, orthocenter of a triangle 163
Contingency 14
Contradiction 14
Contrapositive of a conditional statement 13
Converse of a conditional statement 12

INDEX (continued)

Coordinate plane, properties of triangles and quadrilaterals
    using distance, midpoint, and slope formulas 79-86
Coplanar lines 164-172
Corresponding angles 43-45
Corresponding parts of congruent triangles, CPCT 17, 18, 25-29
Cylinder properties 190-195
Dilations centered at the origin 112-115
Direct Isometry 99, 123
Disjunctions 7-8
Distance Formula 77, 79-86
Distance, invariant 127, 142, 146
Each interior and exterior angle of a regular polygon 47
Equation of a circle, given its center and radius or endpoints of its diameter 88
Equation of a circle, given its graph 89
Equation of a circle in center-radius form, find center and radius 90
Equation of a line, given a point on the line and the equation of a line parallel to the given line 75
    a line perpendicular to the given line 74
Equation of a line that is the perpendicular bisector of a segment
    given the endpoints of the segment 78
Equation of a sphere 201-207
Equations of lines given, find if parallel, perpendicular, or neither 73
Equilateral triangle 79
Exterior angle theorem of a triangle 37-38
Factor of dilation 112
Formal Proofs 25-29
    Congruent triangles 16-18
    Similar triangles 30-31
Glide reflections 107-109
Geometry software to make conjectures 225-230
Graphing circles of the form $(x - h)^2 + (y - k)^2 = r^2$ 91-93
Graphing circles using the graphing calculator 222-223
Graphing systems of equations, one linear and one quadratic 87
HYP-LEG $\cong$ HYP-LEG 16
"IF-AND ONLY-IF" Biconditional 10
"IF-THEN" Conditional 9
Inequality theorem of a triangle 40-42
Inscribed angles 60
Intersecting lines 43-45
Invariant properties under similarities 138-148

INDEX (continued)

Invariant properties under translations, rotations, reflections and glide reflections	107, 112
Inverse of a conditional statement	11
Isometries in the plane	123-126
Isometries, orientation, invariant points, parallelism	99-100, 102
Isometries, identifying specific	127-137
Isosceles right triangle	79
Isosceles trapezoid	80
Isosceles triangle	79
Justify quadrilaterals are parallelograms, rhombuses, rectangles, squares, trapezoids	27-29
Largest angle of a triangle	36, 40-42
Lateral edges of a prism	173
Length of a line segment, given its end points	77
Line perpendicular to a given plane, through a given point	169
Line perpendicular to a given plane	166, 170
Line perpendicular to two intersecting lines	164
Line reflections	102-106
Line segment joining midpoints of two sides of a triangle	32-35
Locus, simple	149-152
Locus, compound	153-154
Locus, graphing compound loci in coordinate plane	155-157
Logic	1-15
Logical equivalent, the contrapositive	14
Longest side of a triangle	36, 40-42
Medians of a triangle	97-98
Midpoint of a line segment, given its endpoints	76, 79-86
Midpoints of two sides of a triangle	50
Negation of a Conditional Statement	3
Negation of a Statement	1-4
Negation of "All" Statements	4
Negation of "Some" Statements	4
Opposite isometry	102, 107, 123
"OR" Disjunction	7-8
Orientation, invariant	127, 142, 146
Parabola	87
Parallel lines in circles	69
Parallelism, invariant	107, 127, 142, 146
Parallelograms	48-49, 80
Parallelograms, involving angles, sides, diagonals	19-20
Perimeter, invariant	127, 142, 146
Perpendicular bisector	102
Perpendicular lines in circles	69
Plane intersecting two parallel planes	171

INDEX (continued)

Plane perpendicular to a given line 165
Planes perpendicular to each other 168
Planes perpendicular to same line 172
Prism, volume formula 176-182
Prisms with equal volumes 174-175
Proof, hypothesis to conclusion 25-32
Proportional relationships of sides of a triangle 32-35, 53
Pythagorean theorem 54-55
Quadrilateral 80
Rectangle 80
Rectangles, involving angles, sides, diagonals 21
Rectangles, proving diagonals are congruent 94-95
Reflections over the lines $x = 0$, $y = 0$, and $y = x$ 102-106
Regular pyramid properties 183-189
Rhombus 80
Rhombuses, involving angles, sides, diagonals 22
Right circular cone properties 196-200
Right triangle 79
Right triangle mean proportionality 51-52
    (Altitude)^2 = segment 1 × segment 2
    (Leg)^2 = hypotenuse × projection
Rotations 110-111
SAS ≅ SAS 16, 138, 146
Scale factor in dilations 112
Scalene triangle 79
Scatterplots, graphing polygons by calculator 215
Scatterplots, rotating polygons about the origin 218-219
Scatterplots, rotating polygons over straight lines 220-221
Scatterplots, transforming polygons 216-217
Segments intersected by a circle 59, 69
    A tangent and a secant from same external point 66
    Two intersecting chords 67
    Two secants from same external point 68
    Two tangents from same external point 70, 71
Similar triangles, corresponding sides are in proportion 30-32
Similar triangles, product of means = product of extremes 30-32
Similarities, identify specific 138-141
Similarities, properties than remain invariant 146-148
Similarities and isometries, properties that remain invariant 142-145
Similarity of triangles, AA, SAS, SSS 30-35
Slope formulas in coordinate plane 79-86
Slope of a line perpendicular to a given equation of a line 72

INDEX (continued)

SOLVER, solving equations using the graphing calculator 213
Solving equations by determining the zeros of a function 214
Sphere properties 201-207
Squares, involving angles, sides, diagonals 23
SSS $\cong$ SSS 16, 138, 146
Statement 14
Sum of measures of angles in polygons 46
Sum of measures of angles of triangles 36
Systems of equations with graphing 87
Tangent line and a chord 61
Tangent line and a secant 63-65
Tangent lines to a circle 58-71
      Common tangents, external and internal 70-71
      Tangent and a chord 61, 64, 65, 69
      Tangent line and radius, form a right angle 61, 69
      Two tangents from same external point 63-65, 70
Tautologies 14-15
Transformational techniques, translations, rotations, reflections 99-148
Translations about the origin of 90 degrees and 180 degrees 99-101
Transversal 43
Trapezoid 80
Trapezoids, involving angles, sides, medians, diagonals 56-57
Trapezoids, isosceles, involving angles, sides, diagonals 24
Truth tables 14, 15
Truth value of a statement 1-2
Two intersecting chords 62
Two lines cut by a transversal 43-45
Two lines perpendicular to the same plane 167
Two parallel lines 73
Two perpendicular lines 73
Two planes perpendicular to the same line 172
Two secants 63-66
Two tangent lines 63-66
Venn Diagram 5, 7

p. 72-98